J. Hamblin (James Hamblin) Smith

Elements of geometry

J. Hamblin (James Hamblin) Smith

Elements of geometry

ISBN/EAN: 9783743481763

Manufactured in Europe, USA, Canada, Australia, Japa

Cover: Foto ©berggeist007 / pixelio.de

Manufactured and distributed by brebook publishing software (www.brebook.com)

J. Hamblin (James Hamblin) Smith

Elements of geometry

𝔚. 𝔍. 𝔊age & 𝔠o.'s 𝔐athematical 𝔚orks.

ELEMENTS OF GEOMETRY:

CONTAINING

BOOKS I. TO VI. AND PORTIONS OF BOOKS
XI. AND XII. OF EUCLID.

WITH EXERCISES AND NOTES

BY

J. HAMBLIN SMITH, M.A.,

OF GONVILLE AND CAIUS COLLEGE, AND LATE LECTURER AT ST.
PETER'S COLLEGE, CAMBRIDGE.

With a Selection of Examination Papers

BY

THOS. KIRKLAND, M. A.,

Principal Normal School, Toronto.

TENTH CANADIAN COPYRIGHT EDITION.

Authorized by the Minister of Education.

W. J. GAGE & COMPANY,

TORONTO.

PREFACE.

To preserve Euclid's order, to supply omissions, to remove defects, to give brief notes of explanation and simpler methods of proof in cases of acknowledged difficulty—such are the main objects of this Edition of the Elements.

The work is based on the Greek text, as it is given in the Editions of August and Peyrard. To the suggestions of the late Professor De Morgan, published in the Companion to the British Almanack for 1849, I have paid constant deference.

A limited use of symbolic representation, wherein the symbols stand for words and not for operations, is generally regarded as desirable, and I have been assured, by the highest authorities on this point, that the symbols employed in this book are admissible in the Examinations at Oxford and Cambridge.[1]

I have generally followed Euclid's method of proof, but not to the exclusion of other methods recom-

---

[1] I regard this point as completely settled in Cambridge by the following notices prefixed to the papers on Euclid set in the Senate-House Examinations :

I. In the Previous Examination :

In answers to these questions any intelligible symbols and abbreviations may be used.

II. In the Mathematical Tripos :

In answers to the questions on Euclid the symbol — must not be used. The only abbreviation admitted for the square on AB is " sq. on AB," and for the rectangle contained by AB and CD, " rect. AB, CD."

mended by their simplicity, such as the demonstrations by which I propose to replace (at least for a first reading) the difficult Theorems 5 and 7 in the First Book. I have also attempted to render many of the proofs, as for instance Propositions 2, 13, and 35 in Book I., and Proposition 13 in Book II., less confusing to the learner.

In Propositions 4, 5, 6, 7, and 8 of the Second Book I have ventured to make an important change in Euclid's mode of exposition, by omitting the diagonals from the diagrams and the gnomons from the text.

In the Third Book I have deviated with even greater boldness from the precise line of Euclid's method. For it is in treating of the properties of the circle that the importance of certain matters, to which reference is made in the Notes of the present volume, is fully brought out. I allude especially to the application of Superposition as a test of equality, to the conception of an Angle as a magnitude capable of unlimited increase, and to the development of the methods connected with Loci and Symmetry.

The Exercises have been selected with considerable care, chiefly from the Senate House Examination Papers. They are intended to be progressive and easy, so that a learner may from the first be induced to work out something for himself.

I desire to express my thanks to the friends who have improved this work by their suggestions, and to beg for further help of the same kind.

J. HAMBLIN SMITH.

CAMBRIDGE, 1872.

CONTENTS.

ELEMENTS OF GEOMETRY.

INTRODUCTORY REMARKS.

WHEN a block of stone is hewn from the rock, we call it a Solid *Body*. The stone-cutter shapes it, and brings it into that which we call *regularity of form ;* and then it becomes a Solid *Figure*.

Now suppose the figure to be such that the block has six flat sides, each the exact counterpart of the others ; so that, to one who stands facing a corner of the block, the three sides which are visible present the appearance represented in this diagram.

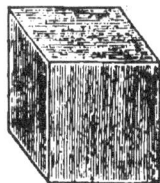

Each side of the figure is called a *Surface ;* and when smoothed and polished, it is called a *Plane* Surface.

The sharp and well-defined edges, in which each pair of sides meets, are called *Lines*.

The place, at which any three of the edges meet, is called a *Point*.

A *Magnitude* is anything which is made up of parts in any way like itself. Thus, a line is a magnitude ; because we may regard it as made up of parts which are themselves lines.

The properties Length, Breadth (or Width), and Thickness (or Depth or Height) of a body are called its *Dimensions.*

We make the following distinction between Solids, Surfaces, Lines, and Points :

A Solid has three dimensions, Length, Breadth, Thickness.

A Surface has two dimensions, Length, Breadth.

A Line has one dimension, Length.

A point has no dimensions.

S. E.

1

BOOK I.

DEFINITIONS.

I. A POINT is that which has no parts.

This is equivalent to saying that a Point has no magnitude, since we define it as that which cannot be divided into smaller parts.

II. A LINE is length without breadth.

We cannot conceive a visible line without breadth; but we can reason about lines as if they had no breadth, and this is what Euclid requires us to do.

III. The EXTREMITIES of finite LINES are points.

A point marks *position*, as for instance, the place where a line begins or ends, or meets or crosses another line.

IV. A STRAIGHT LINE is one which lies in the same direction from point to point throughout its length.

V. A SURFACE is that which has length and breadth only.

VI. The EXTREMITIES of a SURFACE are lines.

VII. A PLANE SURFACE is one in which, if any two points be taken, the straight line between them lies wholly in that surface.

Thus the ends of an uncut cedar-pencil are plane surfaces; but the rest of the surface of the pencil is not a plane surface, since two points may be taken in it such that the *straight* line joining them will not lie on the surface of the pencil.

In our introductory remarks we gave examples of a Surface, a Line, and a Point, as we know them through the evidence of the senses.

The Surfaces, Lines, and Points of Geometry may be regarded as mental pictures of the surfaces, lines, and points which we know from experience.

It is, however, to be observed that Geometry requires us to conceive the possibility of the existence
        of a Surface apart from a Solid body,
        of a Line apart from a Surface.
        of a Point apart from a Line.

VIII. When two straight lines meet one another, the inclination of the lines to one another is called an ANGLE.

When *two* straight lines have one point common to both, they are said to *form* an angle (or angles) at that point. The point is called the *vertex* of the angle (or angles), and the lines are called the *arms* of the angle (or angles).

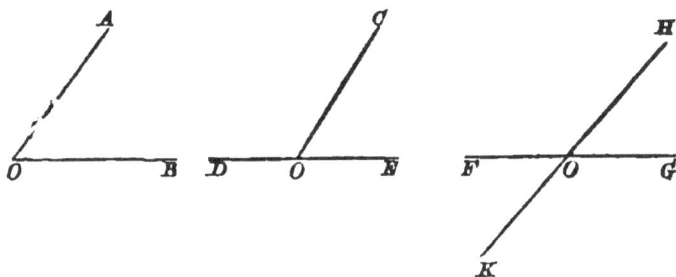

Thus, if the lines *OA*, *OB* are terminated at the same point *O*, they form an angle, which is called *the angle at O*, or *the angle AOB*, or *the angle BOA*,—the letter which marks the vertex being put between those that mark the arms.

Again, if the line *CO* meets the line *DE* at a point in the line *DE*, so that *O* is a point common to both lines, *CO* is said to make with *DE* the angles *COD*, *COE* ; and these (as having one arm, *CO*, common to both) are called *adjacent* angles.

Lastly, if the lines *FG*, *HK* cut each other in the point *O*, the lines make with each other four angles *FOH*, *HOG*, *GOK*, *KOF*; and of these *GOH*, *FOK* are called *vertically opposite* angles, as also are *FOH* and *GOK*.

When *three or more* straight lines as OA, OB, OC, OD have a point O common to all, the angle formed by one of them, OD,

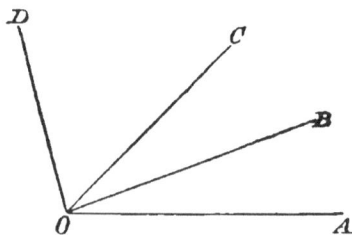

with OA may be regarded as being made up of the angles AOD, BOC, COD; that is, we may speak of the angle AOD as a whole, of which the parts are the angles AOB, BOC, and COD.

Hence we may regard an angle as a *Magnitude*, inasmuch as any angle may be regarded as being made up of parts which are themselves angles.

The size of an angle depends in no way on the length of the arms by which it is bounded.

We shall explain hereafter the restriction on the magnitude of angles enforced by Euclid's definition, and the important results that follow an extension of the definition.

IX. When a straight line (as AB) meeting another straight line (as CD) makes the adjacent angles (ABC and ABD) equal to one another, each of the angles is called a RIGHT ANGLE; and each line is said to be a PER- PENDICULAR to the other.

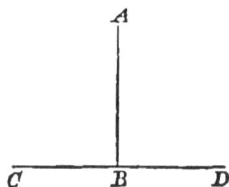

X. An OBTUSE ANGLE is one which is greater than a right angle.

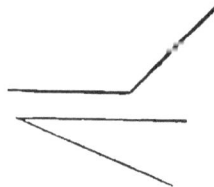

XI. An ACUTE ANGLE is one which is less than a right angle.

XII. A FIGURE is that which is enclosed by one or more boundaries.

XIII. A CIRCLE is a plane figure contained by one line, which is called the CIRCUMFERENCE, and is such, that all straight lines drawn to the circumference from a certain point (called the CENTRE) within the figure are equal to one another.

XIV. Any straight line drawn from the centre of a circle to the circumference is called a RADIUS.

XV. A DIAMETER of a circle is a straight line drawn through the centre and terminated both ways by the circumference.

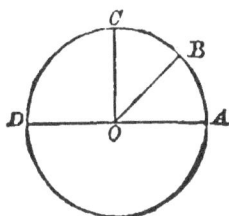

Thus, in the diagram, O is the centre of the circle $ABCD$, OA, OB, OC, OD are Radii of the circle, and the straight line AOD is a Diameter. Hence the radius of a circle is half the diameter.

XVI. A SEMICIRCLE is the figure contained by a diameter and the part of the circumference cut off by the diameter.

XVII. RECTILINEAR figures are those which are contained by straight lines.

The PERIMETER (or Periphery) of a rectilinear figure is the sum of its sides.

XVIII. A TRIANGLE is a plane figure contained by three straight lines.

XIX. A QUADRILATERAL is a plane figure contained by four straight lines.

XX. A POLYGON is a plane figure contained by more than four straight lines.

When a polygon has all its sides equal and all its angles equal it is called a *regular* polygon.

XXI. An EQUILATERAL Triangle is one which has all its sides equal.

XXII. An ISOSCELES Triangle is one which has two sides equal.

The third side is often called the *base* of the triangle.

The term *base* is applied to any one of the sides of a triangle to distinguish it from the other two, especially when they have been previously mentioned.

XXIII. A RIGHT-ANGLED Triangle is one in which one of the angles is a right angle.

The side *subtending*, that is, *which is opposite* the right angle, is called the *Hypotenuse*.

XXIV. An OBTUSE-ANGLED Triangle is one in which one of the angles is obtuse.

It will be shewn hereafter that a triangle can have only one of its angles either equal to, or greater than, a right angle.

XXV. An ACUTE-ANGLED Triangle is one in which ALL the angles are acute.

XXVI. PARALLEL STRAIGHT LINES are such as, being in the same plane, never meet when continually produced in both directions.

Euclid proceeds to put forward Six Postulates, or Requests, that he may be allowed to make certain assumptions on the construction of figures and the properties of geometrical magnitudes.

POSTULATES

Let it be granted—

I. That a straight line may be drawn from any one point to any other point.

II. That a terminated straight line may be produced to any length in a straight line.

III. That a circle may be described from any centre at any distance from that centre.

IV. That all right angles are equal to one another.

V. That two straight lines cannot enclose a space.

VI. That if a straight line meet two other straight lines, so as to make the two interior angles on the same side of it, taken together, less than two right angles, these straight lines being continually produced shall at length meet upon that side, on which are the angles, which are together less than two right angles.

The word rendered "Postulates" is in the original αἰτήματα, "requests."

In the first three Postulates Euclid states the use, under certain restrictions, which he desires to make of certain instruments for the construction of lines and circles.

In Post. I. and II. he asks for the use of the straight ruler, wherewith to draw straight lines. The restriction is, that the ruler is not supposed to be marked with divisions so as to measure lines.

In Post. III. he asks for the use of a pair of compasses, wherewith to describe a circle, whose centre is at one extremity of a given line, and whose circumference passes through the other extremity of that line. The restriction is, that the compasses are not supposed to be capable of conveying distances.

Post. IV. and V. refer to simple geometrical facts, which Euclid desires to take for granted.

Post. VI. may, as we shall shew hereafter, be deduced from a more simple Postulate. The student must defer the consideration of this Postulate, till he has reached the 17th Proposition of Book I.

Euclid next enumerates, as statements of fact, nine Axioms

or, as he calls them, Common Notions, applicable (with the exception of the eighth) to all kinds of magnitudes, and not necessarily restricted, as are the Postulates, to *geometrical magnitudes.*

AXIOMS.

I. Things which are equal to the same thing are equal to one another.

II. If equals be added to equals, the wholes are equal.

III. If equals be taken from equals, the remainders are equal.

IV. If equals and unequals be added together, the wholes are unequal.

V. If equals be taken from unequals, or unequals from equals, the remainders are unequal.

VI. Things which are double of the same thing, or of equal things, are equal to one another.

VII. Things which are halves of the same thing, or of equal things, are equal to one another.

VIII. Magnitudes which coincide with one another are equal to one another.

IX. The whole is greater than its part.

With his Common Notions Euclid takes the ground of authority, saying in effect, " To my Postulates I request, to my Common Notions I claim, your assent."

Euclid develops the science of Geometry in a series of Propositions, some of which are called Theorems and the rest Problems, though Euclid himself makes no such distinction.

By the name *Theorem* we understand a truth, capable of demonstration or proof by deduction from truths previously admitted or proved.

By the name *Problem* we understand a construction, capable of being effected by the employment of principles of construction previously admitted or proved.

A *Corollary* is a Theorem or Problem easily deduced from, or effected by means of, a Proposition to which it is attached.

We shall divide the First Book of the Elements into three sections. The reason for this division will appear in the course of the work.

SYMBOLS AND ABBREVIATIONS USED IN BOOK I.

∵	*for*	because	⊙	*for*	circle
∴	............therefore		○ce	............circumference
=	............is (or are) equal to		‖	...............parallel
∠	............angle		▱	............parallelogram
△	.........triangle		⊥	............perpendicular

equilat. .....equilateral		reqd. ....required
extr..........exterior		rt...............right
intr..........interior		sq. ....square
pt............point		sqq............squares
rectil. ......rectilinear		st...............straight

It is well known that one of the chief difficulties with learners of Euclid is to distinguish between what is assumed, or given, and what has to be proved in some of the Propositions. To make the distinction clearer we shall put in italics the statements of what has to be done in a Problem, and what has to be proved in a Theorem. The last line in the proof of every Proposition states, that what had to be done or proved has been done or proved.

The letters Q. E. F. at the end of a Problem stand for *Quod erat faciendum.*

The letters Q. E. D. at the end of a Theorem stand for *Quod erat demonstrandum.*

In the marginal references :
Post. stands for Postulate.
Def. .... Definition.
Ax. .... Axiom.
I. 1. .... Book I. Proposition 1.

Hyp. stands for Hypothesis, *supposition*, and refers to something granted, or assumed to be true.

SECTION I.

On the Properties of Triangles.

PROPOSITION I. PROBLEM.

To describe an equilateral triangle on a given straight line.

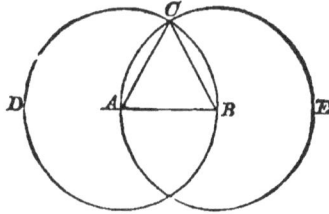

Let AB be the given st. line.

It is required to describe an equilat. △ *on* AB

With centre A and distance AB describe ⊙ BCD.   Post. 3.

With centre B and distance BA describe ⊙ ACE.   Post. 3.

From the pt. C, in which the ⊙s cut one another,

    draw the st. lines CA, CB.   Post. 1.

Then will ABC be an equilat. △.

For   ∵ A is the centre of ⊙ BCD,

        ∴ $AC = AB$.   Def. 13.

And   ∵ B is the centre of ⊙ ACE,

        ∴ $BC = AB$.   Def. 13.

Now   ∵ AC, BC are each $= AB$,

        ∴ $AC = BC$.   Ax. 1.

Thus AC, AB, BC are all equal, and an equilat. △ ABC has been described on AB.

                                  Q. E. F.

From a given point to draw a straight line equal to a given straight line.

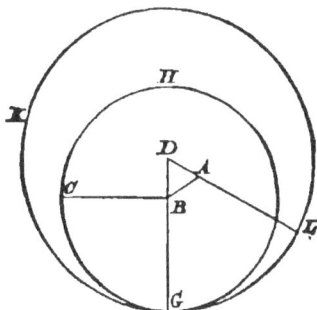

Let A be the given pt., and BC the given st. line.

It is required to draw from A a st. line equal to BC.

From A to B draw the st. line AB.	Post. 1.
On AB describe the equilat. $\triangle$ ABD.	I. 1.
With centre B and distance BC describe $\odot$ CGH.	Post. 3.
Produce DB to meet the $\bigcirc$cc CGH in G.	
With centre D and distance DG describe $\odot$ GKL.	Post. 3.
Produce DA to meet the $\bigcirc$cc GKL in L.	
Then will $AL=BC$.	

For	$\because$ B is the centre of $\odot$ CGH,	
	$\therefore$ $BC=BG$.	Def. 13.
And	$\because$ D is the centre of $\odot$ GKL,	
	$\therefore$ $DL=DG$.	Def. 13.
And parts of these, DA and DB, are equal.		Def. 21.
	$\therefore$ remainder $AL=$ remainder BG.	Ax. 3.
	But $BC=BG$;	
	$\therefore$ $AL=BC$.	Ax. 1.

Thus from pt. A a st. line AL has been drawn $=BC$.

Q. E. F.

PROPOSITION III. PROBLEM.

From the greater of two given straight lines to cut off a part equal to the less.

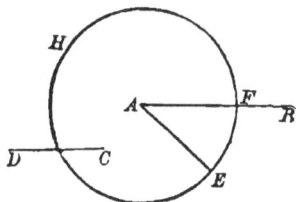

Let AB be the greater of the two given st. lines AB, CD.

It is required to cut off from AB a part $= CD$.

From A draw the st. line $AE = CD$.   I. 2.

With centre A and distance AE describe $\odot$ EFH,
    cutting AB in F.

Then will $AF = CD$.

For $\because$ A is the centre of $\odot$ EFH,

                      $\therefore$ $AF = AE$.

But $AE = CD$;

                      $\therefore$ $AF = CD$.   Ax. 1.

Thus from AB a part AF has been cut off $= CD$.

                                        Q. E. F.

EXERCISES.

1. Shew that if straight lines be drawn from A and B in the diagram of Prop. I. to the other point in which the circles intersect, another equilateral triangle will be described on AB.

2. By a construction similar to that in Prop. III. produce the less of two given straight lines that it may be equal to the greater.

3. Draw a figure for the case in Prop. II., in which the given point coincides with B.

4. By a similar construction to that in Prop. I. describe on a given straight line an isosceles triangle, whose equal sides shall be each equal to another given straight line.

If two triangles have two sides of the one equal to two sides of the other, each to each, and have likewise the angles contained by those sides equal to one another, they must have their third sides equal ; and the two triangles must be equal, and the other angles must be equal, each to each, viz. those to which the equal sides are opposite.

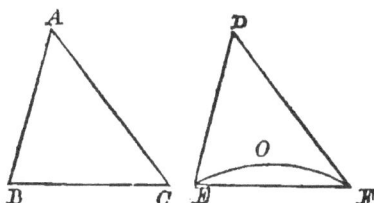

In the △ s *ABC, DEF*,
let *AB=DE*, and *AC=DF*, and ∠ *BAC*= ∠ *EDF*.

Then must BC=EF and △ ABC = △ DEF, and the other ∠ s, to which the equal sides are opposite, must be equal, that is, ∠ ABC= ∠ DEF and ∠ ACB= ∠ DFE.

For, if △ *ABC* be applied to △ *DEF*,
so that *A* coincides with *D*, and *AB* falls on *DE*,
then ∵ *AB=DE*, ∴ *B* will coincide with *E*.

And ∵ *AB* coincides with *DE*, and ∠ *BAC*= ∠ *EDF*, Hyp.
∴ *AC* will fall on *DF*.

Then ∵ *AC=DF*, ∴ *C* will coincide with *F*.
And ∵ *B* will coincide with *E*, and *C* with *F*,
∴ *BC* will coincide with *EF* ;

for if not, let it fall otherwise as *EOF* : then the two st. lines *BC, EF* will enclose a space, which is impossible.   Post. 5.

∴ *BC* will coincide with and ∴ is equal to *EF*,   Ax. 8.

and △ *ABC*.. △ *DEF*,

and ∠ *ABC*.. ∠ *DEF*,

and ∠ *ACB*.. ∠ *DFE*.

Q. E. D.

NOTE 1. *On the Method of Superposition.*

Two geometrical magnitudes are said, in accordance with Ax. VIII. to be *equal*, when they can be so placed that the boundaries of the one coincide with the boundaries of the other.

Thus, two straight lines are equal, if they can be so placed that the points at their extremities coincide : and two angles are equal, if they can be so placed that their vertices coincide in position and their arms in direction : and two triangles are equal, if they can be so placed that their sides coincide in direction and magnitude.

In the application of the test of equality by this *Method of Superposition*, we assume that an angle or a triangle may be moved from one place, turned over, and put down in another place, without altering the relative positions of its boundaries.

We also assume that if one part of a straight line coincide with one part of another straight line, the other parts of the lines also coincide in direction ; or, that straight lines, which coincide in two points, coincide when produced.

The method of Superposition enables us also to compare magnitudes of the same kind that are unequal. For example, suppose *ABC* and *DEF* to be two given angles.

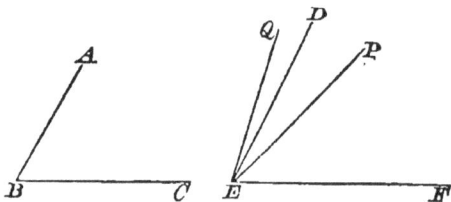

Suppose the arm *BC* to be placed on the arm *EF*, and the vertex *B* on the vertex *E*.

Then, if the arm *BA* coincide in direction with the arm *ED*, the angle *ABC* is equal to *DEF*.

If *BA* fall between *ED* and *EF* in the direction *EP*, *ABC* is less than *DEF*.

If *BA* fall in the direction *EQ* so that *ED* is between *EQ* and *EF*, *ABC* is greater than *DEF*.

NOTE 2. *On the Conditions of Equality of two Triangles.*

A Triangle is composed of six parts, three sides and three angles.

When the six parts of one triangle are equal to the six parts of another triangle, each to each, the Triangles are said to be equal in all respects.

There are four cases in which Euclid proves that two triangles are equal in all respects ; viz., when the following parts are equal in the two triangles.

1. Two sides and the angle between them.   I. 4.

2. Two angles and the side between them.   I. 26.

3. The three sides of each.   I. 8.

4. Two angles and the side opposite one of them.   I. 26.

The Propositions, in which these cases are proved, are the most important in our First Section.

The first case we have proved in Prop. IV.

Availing ourselves of the method of superposition, we can prove Cases 2 and 3 by a process more simple than that employed by Euclid, and with the further advantage of bringing them into closer connexion with Case 1. We shall therefore give three Propositions, which we designate A, B, and C, in the Place of Euclid's Props. V. VI. VII. VIII.

The displaced Propositions will be found on pp. 108-112.

Proposition A corresponds with Euclid I. 5.

.... B .... I. 26, first part.

.... C .... I. 8.

PROPOSITION A. THEOREM.

*If two sides of a triangle be equal, the angles opposite those
sides must also be equal.*

FIG. 1.　　　FIG. 2.

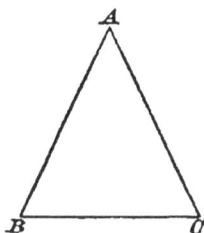 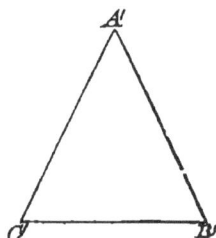

In the isosceles triangle ABC, let $AC=AB$.　(Fig. 1.)

Then must $\angle ABC = \angle ACB$.

Imagine the $\triangle ABC$ to be taken up, turned round, and set
down again in a reversed position as in Fig. 2, and designate
the angular points A', B', C'.

Then in $\triangle$s ABC, $A'C'B'$,

∵ $AB=A'C'$, and $AC=A'B'$, and $\angle BAC = \angle C'A'B'$,

∴ $\angle ABC = \angle A'C'B'$.　　　　　I. 4.

But　　　　$\angle A'C'B' = \angle ACB$;

∴ $\angle ABC = \angle ACB$.　　　　Ax. 1.

Q.E.D.

COR. Hence every equilateral triangle is also equiangular.

NOTE. When one side of a triangle is distinguished from
the other sides by being called the *Base*, the angular point opposite to that side is called the *Vertex* of the triangle.

PROPOSITION B. THEOREM.

If two triangles have two angles of the one equal to two angles of the other, each to each, and the sides adjacent to the equal angles in each also equal ; then must the triangles be equal in all respects.

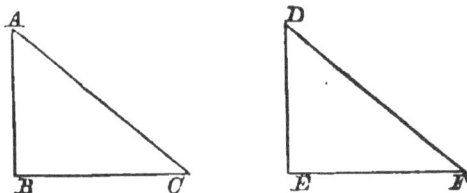

In △s *ABC, DEF,*

let ∠ *ABC* = ∠ *DEF*, and ∠ *ACB* = ∠ *DFE*, and *BC* = *EF*.

Then must *AB* = *DE*, and *AC* = *DF*, and ∠ *BAC* = ∠ *EDF*.

For if △ *DEF* be applied to △ *ABC*, so that *E* coincides with *B*, and *EF* falls on *BC* ;

then ∵ *EF* = *BC*, ∴ *F* will coincide with *C* ;

and ∵ ∠ *DEF* = ∠ *ABC*, ∴ *ED* will fall on *BA* ;

∴ *D* will fall on *BA* or *BA* produced.

Again, ∵ ∠ *DFE* = ∠ *ACB*, ∴ *FD* will fall on *CA* ;

∴ *D* will fall on *CA* or *CA* produced.

∴ *D* must coincide with *A*, the only pt. common to *BA* and *CA*.

∴ *DE* will coincide with and ∴ is equal to *AB*,

and *DF* .... *AC*,

and ∠ *EDF* .... ∠ *BAC*,

and △ *DEF* .... △ *ABC* ;

and ∴ the triangles are equal in all respects.

Q. E. D.

COR. Hence, by a process like that in Prop. A, we can prove the following theorem :

If two angles of a triangle be equal, the sides which subtend them are also equal. (Eucl. I. 6.)

Proposition C.   Theorem.

If two triangles have the three sides of the one equal to the three sides of the other, each to each, the triangles must be equal in all respects.

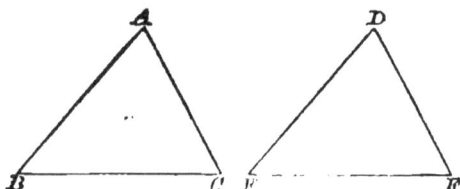

Let the three sides of the △s *ABC, DEF* be equal, each to each, that is, $AB=DE$, $AC=DF$, and $BC=EF$.

Then must the triangles be equal in all respects.

Imagine the △ *DEF* to be turned over and applied to the △ *ABC*, in such a way that *EF* coincides with *BC*, and the vertex *D* falls on the side of *BC* opposite to the side on which *A* falls ; and join *AD*.

CASE I.   When *AD* passes through *BC*.

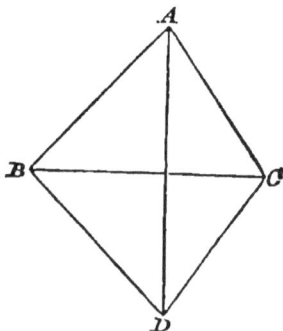

Then in △ *ABD*, ∵ *BD=BA*, ∴ ∠ *BAD*= ∠ *BDA*, I. A.
And in △ *ACD*, ∵ *CD=CA*, ∴ ∠ *CAD*= ∠ *CDA*, I. A.
∴ sum of ∠s *BAD, CAD*=sum of ∠s *BDA, CDA*,   Ax. 2.
that is,   ∠ *BAC*= ∠ *BDC*.

Hence we see, referring to the original triangles, that
                ∠ *BAC*= ∠ *EDF*.

∴ by Prop. 4, the triangles are equal in all respects.

CASE II. When the line joining the vertices does not pass through *BC.*

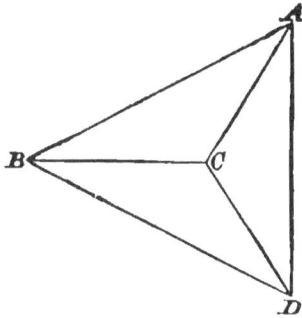

Then in △ *ABD*, ∵ *BD=BA*, ∴ ∠ *BAD=* ∠ *BDA*,   I. A.

And in △ *ACD*, ∵ *CD=CA*, ∴ ∠ *CAD=* ∠ *CDA*,   I. A.

Hence since the whole angles *BAD, BDA* are equal.

and parts of these *CAD, CDA* are equal.

∴ the remainders *BAC, BDC* are equal.   Ax. 3.

Then, as in Case I., the equality of the original triangles may be proved.

CASE III. When *AC* and *CD* are in the same straight line.

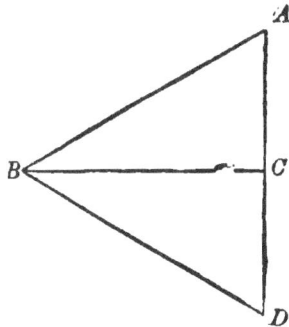

Then in △ *ABD*, ∵ *BD=BA*, ∴ ∠ *BAD=* ∠ *BDA*,   I. A.

that is, ∠ *BAC=* ∠ *BDC*.

Then, as in Case I., the equality of the original triangles may be proved.

Q. E. D.

PROPOSITION IX. PROBLEM.

To bisect a given angle.

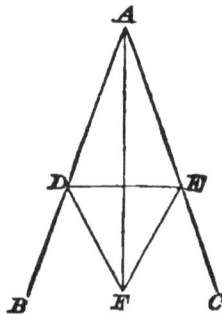

Let BAC be the given angle.

It is required to bisect $\angle BAC$.

In AB take any pt. D.

In AC make $AE = AD$, and join DE.

On DE, on the side remote from A, describe an equilat. $\triangle DFE$.   I. 1.

Join AF. Then AF will bisect $\angle BAC$.

For in $\triangle$ s AFD, AFE,

$\because AD = AE$, and AF is common, and $FD = FE$,

$\therefore \angle DAF = \angle EAF$,   I. c.

that is, $\angle BAC$ is bisected by AF.

Q. E. F.

Ex. 1. Shew that we can prove this Proposition by means of Prop. IV, and Prop. A , without applying Prop. C.

Ex. 2. If the equilateral triangle, employed in the construction, be described with its vertex towards the given angle ; shew that there is one case in which the construction will fail, and two in which it will hold good.

NOTE.—The line dividing an angle into two equal parts is called the BISECTOR of the angle.

PROPOSITION X. PROBLEM.

To bisect a given finite straight line.

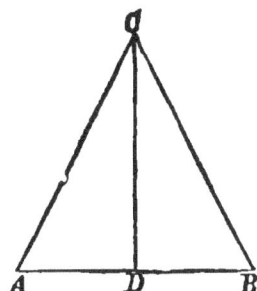

Let *AB* be the given st. line.

It is required to bisect AB.

On *AB* describe an equilat. △ *ACB*.   I. 1.

Bisect ∠ *ACB* by the st. line *CD* meeting *AB* in *D* ;   ∠ 9.
then *AB* shall be bisected in *D*.

For in △ s *ACD*, *BCD*,

∵ *AC*=*BC*, and *CD* is common, and ∠ *ACD*= ∠ *BCD*,

∴ *AD*=*BD* ;   I. 4.

∴ *AB* is bisected in *D*.

Q. E. F.

Ex. 1. The straight line, drawn to bisect the vertical angle of an isosceles triangle, also bisects the base.

Ex. 2. The straight line, drawn from the vertex of an isosceles triangle to bisect the base, also bisects the vertical angle.

Ex. 3. Produce a given finite straight line to a point, such that the part produced may be one-third of the line, which is made up of the whole and the part produced.

PROPOSITION XI. PROBLEM.

To draw a straight line at right angles to a given straight line from a given point in the same.

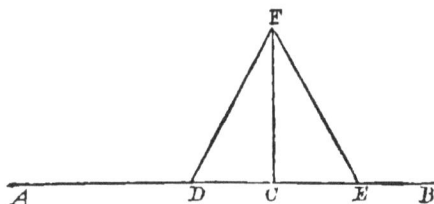

Let AB be the given st. line, and C a given pt. in it.

It is required to draw from C a st. line $\perp$ to AB.

Take any pt. D in AC, and in CB make $CE = CD$.

On DE describe an equilat. $\triangle$ DFE.   I. 1.

Join FC. FC shall be $\perp$ to AB.

For in $\triangle$s DCF, ECF,

    $\because DC = CE$, and CF is common, and $FD = FE$,

      $\therefore \angle DCF = \angle ECF$;   I. c.

     and $\therefore FC$ is $\perp$ to AB.   Def. 9.

Q. E. F.

COR. To draw a straight line at right angles to a given straight line AC from one extremity, C, take any point D in AC, produce AC to E, making $CE = CD$, and proceed as in the proposition.

Ex. 1. Show that in the diagram of Prop. IX. AF and ED intersect each other at right angles, and that ED is bisected by AF.

Ex. 2. If O be the point in which two lines, bisecting AB and AC, two sides of an equilateral triangle, at right angles, meet; shew that OA, OB, OC are all equal.

Ex. 3. Shew that Prop. XI. is a particular case of Prop. IX.

To draw a straight line perpendicular to a given straight line of an unlimited length from a given point without it.

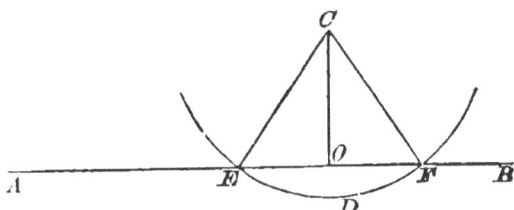

Let AB be the given st. line of unlimited length; C the given pt. without it.

It is required to draw from C a st. line $\perp$ to AB.

Take any pt. D on the other side of AB.

With centre C and distance CD describe a $\odot$ cutting AB in E and F.

Bisect EF in O, and join CE, CO, CF.   I. 10

Then CO shall be $\perp$ to AB.

For in △ s COE, COF,

∵ $EO = FO$, and CO is common, and $CE = CF$,

∴ $\angle COE = \angle COF$;   I. c.

∴ CO is $\perp$ to AB.   Def. 9.

Q. E. F.

Ex. 1. If the straight line were not of unlimited length, how might the construction fail?

Ex. 2. If in a triangle the perpendicular from the vertex on the base bisect the base, the triangle is isosceles.

Ex. 3. The lines drawn from the angular points of an equilateral triangle to the middle points of the opposite sides are equal.

Miscellaneous Exercises on Props. I. to XII.

1. Draw a figure for Prop. II. for the case when the given point A is

    (a) below the line BC and to the right of it.

    (β) below the line BC and to the left of it.

2. Divide a given angle into four equal parts.

3. The angles B, C, at the base of an isosceles triangle, are bisected by the straight lines BD, CD, meeting in D; shew that BDC is an isosceles triangle.

4. D, E, F are points taken in the sides BC, CA, AB, of an equilateral triangle, so that $BD=CE=AF$. Shew that the triangle DEF is equilateral.

5. In a given straight line find a point equidistant from two given points; 1st, on the same side of it; 2d, on opposite sides of it.

6. ABC is a triangle having the angle ABC acute. In BA, or BA produced, find a point D such that $BD=CD$.

7. The equal sides AB, AC, of an isosceles triangle ABC are produced to points F and G, so that $AF=AG$. BG and CF are joined, and H is the point of their intersection. Prove that $BH=CH$, and also that the angle at A is bisected by AH.

8. BAC, BDC are isosceles triangles, standing on opposite sides of the same base BC. Prove that the straight line from A to D bisects BC at right angles.

9. In how many directions may the line AE be drawn in Prop. III.?

10. The two sides of a triangle being produced, if the angles on the other side of the base be equal, shew that the triangle is isosceles.

11. ABC, ABD are two triangles on the same base AB and on the same side of it, the vertex of each triangle being outside the other. If $AC=AD$, shew that BC cannot $=BD$.

12. From C any point in a straight line AB, CD is drawn at right angles to AB, meeting a circle described with centre A and distance AB in D; and from AD, AE is cut off $=AC$: shew that AEB is a right angle.

PROPOSITION XIII. THEOREM.

The angles which one straight line makes with another upon one side of it are either two right angles, or together equal to two right angles.

Fig. 1.   Fig. 2.

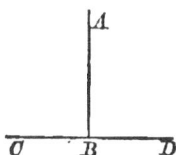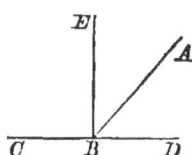

Let AB make with CD upon one side of it the ∠s ABC, ABD.

Then must these be either two rt. ∠ s,
or together equal to two rt. ∠ s

First, if ∠ $ABC =$ ∠ ABD as in Fig. 1,

each of them is a rt. ∠ .   Def. 9.

Secondly, if ∠ ABC be not $=$ ∠ ABD, as in Fig. 2,

from B draw BE ⊥ to CD.   I. 11.

Then sum of ∠s ABC, $ABD =$ sum of ∠s EBC, EBA, ABD,

and sum of ∠s EBC, $EBD =$ sum of ∠s EBC, EBA, ABD ;

∴ sum of ∠s ABC, $ABD =$ sum of ∠s EBC, EBD ;

Ax. 1.

∴ sum of ∠s ABC, $ABD =$ sum of a rt. ∠ and a rt. ∠ ;

∴ ∠s ABC, ABD are together $=$ two rt. ∠ s.

Q. E. D.

Ex. Straight lines drawn connecting the opposite angular points of a quadrilateral figure intersect each other in O. Shew that the angles at O are together equal to four right angles.

NOTE (1.) If two angles together make up a right angle, each is called the COMPLEMENT of the other. Thus, in fig. 2. ∠ ABD is the complement of ∠ ABE.

(2.) If two angles together make up two right angles, each is called the SUPPLEMENT of the other. Thus, in both figures, ∠ ABD is the supplement of ∠ ABC.

PROPOSITION XIV.   THEOREM

If, at a point in a straight line, two other straight lines, upon the opposite sides of it, make the adjacent angles together equal to two right angles, these two straight lines must be in one and the same straight line.

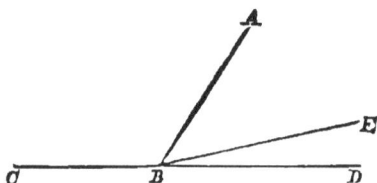

At the pt. B in the st. line AB let the st. lines BC, BD, on opposite sides of AB, make $\angle$ s ABC, ABD together $=$ two rt. angles.

Then BD must be in the same st. line with BC.

For if not, let BE be in the same st. line with BC.

Then $\angle$ s ABC, ABE together $=$ two rt. $\angle$ s.   I. 13.

And $\angle$ s ABC, ABD together $=$ two rt. $\angle$ s.   Hyp.

$\therefore$ sum of $\angle$ s ABC, ABE $=$ sum of $\angle$ s ABC, ABD.

Take away from each of these equals the $\angle$ ABC ;

then $\angle$ $ABE = \angle$ ABD,   Ax. 3.

that is, the less $=$ the greater ; which is impossible,

$\therefore$ BE is not in the same st. line with BC.

Similarly it may be shewn that no other line but BD is in the same st. line with BC.

$\therefore$ BD is in the same st. line with BC.

Q. E. D.

Ex. Shew the necessity of the words *the opposite sides* in the enunciation.

If two straight lines cut one another, the vertically opposite angles must be equal.

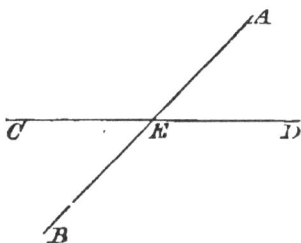

Let the st. lines *AB, CD* cut one another in the pt. *E.*

Then must ∠ *AEC* = ∠ *BED* and ∠ *AED* = ∠ *BEC.*

For ∵ *AE* meets *CD*,

∴ sum of ∠ s *AEC, AED* = two rt. ∠ s.   I. 13.

And ∵ *DE* meets *AB*,

∴ sum of ∠ s *BED, AED* = two rt. ∠ s ;   I. 13.

∴ sum of ∠ s *AEC, AED* = sum of ∠ s *BED, AED* ;

∴ ∠ *AEC* = ∠ *BED*.   Ax. 3.

Similarly it may be shewn that ∠ *AED* = ∠ *BEC.*

Q. E. D.

COROLLARY I. From this it is manifest, that if two straight lines cut one another, the four angles, which they make at the point of intersection, are together equal to four right angles.

COROLLARY II. All the angles, made by any number of straight lines meeting in one point, are together equal to four right angles.

Ex. 1. Shew that the bisectors of *AED* and *BEC* are in the same straight line.

Ex. 2. Prove that ∠ *AED* is equal to the angle between two straight lines drawn at right angles from *E* to *AE* and *EC*, if both lie above *CD*.

Ex. 3. If *AB, CD* bisect each other in *E* ; shew that the triangles *AED, BEC* are equal in all respects.

Note 3. On Euclid's definition of an Angle.

Euclid directs us to regard an angle as the inclination of two straight lines to each other, which meet, *but are not in the same straight line.*

Thus he does not recognise the existence of a single angle equal in magnitude to two right angles.

The words printed in italics are omitted as needless, in Def. viii., p. 3, and that definition may be extended with advantage in the following terms :—

Def. Let *WQE* be a fixed straight line, and *QP* a line which revolves about the fixed point *Q*, and which at first coincides with *QE*.

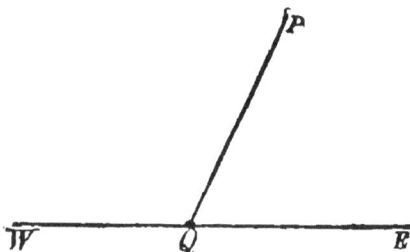

Then, when *QP* has reached the position represented in the diagram, we say that it has described the angle *EQP*.

When *QP* has revolved so far as to coincide with *QW*, we say that it has described an angle *equal to two right angles.*

Hence we may obtain an easy proof of Prop. xiii. ; for whatever the position of *PQ* may be, the angles which it makes with *WE* are together equal to two right angles.

Again, in Prop. xv. it is evident that ∠*AED* = ∠*BEC*, since each has the same supplementary ∠*AEC*.

We shall shew hereafter, p. 149, how this definition may be extended, so as to embrace angles *greater than two right angles.*

If one side of a triangle be produced, the exterior angle is greater than either of the interior opposite angles.

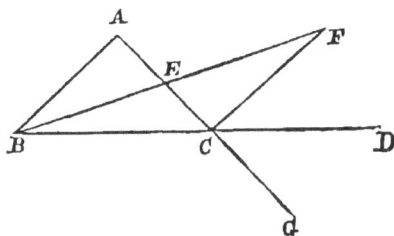

Let the side BC of $\triangle ABC$ be produced to D.

Then must $\angle ACD$ be greater than either $\angle CAB$ or $\angle ABC$.

Bisect AC in E, and join BE.   I. 10.

Produce BE to F, making $EF = BE$, and join FC.

Then in $\triangle$s BEA, FEC,

   $\because BE = FE$, and $EA = EC$, and $\angle BEA = \angle FEC$,   I. 15.

        $\therefore \angle ECF = \angle EAB$.   I. 4.

Now $\angle ACD$ is greater than $\angle ECF$;   Ax. 9.

     $\therefore \angle ACD$ is greater than $\angle EAB$,

that is, $\angle ACD$ is greater than $\angle CAB$.

Similarly, if AC be produced to G it may be shewn that

        $\angle BCG$ is greater than $\angle ABC$.

and $\angle BCG = \angle ACD$;   I. 15.

     $\therefore \angle ACD$ is greater than $\angle ABC$.

                       Q. E. D.

Ex. 1. From the same point there cannot be drawn more than two equal straight lines to meet a given straight line.

Ex. 2. If, from any point, a straight line be drawn to a given straight line making with it an acute and an obtuse angle, and if, from the same point, a perpendicular be drawn to the given line ; the perpendicular will fall on the side of the acute angle.

PROPOSITION XVII. THEOREM.

Any two angles of a triangle are together less than two right angles.

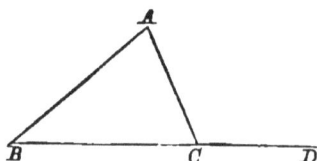

Let ABC be any $\triangle$.

Then must any two of its $\angle$ s be together less than two rt. $\angle$ s.

Produce BC to D.

Then $\angle ACD$ is greater than $\angle ABC$.   I. 16.

∴ $\angle$ s ACD, ACB are together greater than $\angle$ s ABC, ACB.

But $\angle$ s ACD, ACB together = two rt. $\angle$ s.   I. 13.

∴ $\angle$ s ABC, ACB are together less than two rt. $\angle$ s.

Similarly it may be shewn that $\angle$ s ABC, BAC and also that $\angle$ s BAC, ACB are together less than two rt. $\angle$ s.

Q. E. D

NOTE 4. *On the Sixth Postulate.*

We learn from Prop. XVII. that if two straight lines BM and CN, which meet in A, are met by another straight line DE in the points O, P,

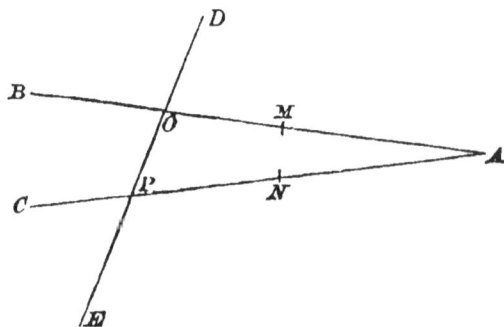

the angles MOP and NPO are together less than two right angles.

The Sixth Postulate asserts that if a line DE meeting two other lines BM, CN makes MOP, NPO, the two interior angles on the same side of it, together less than two right angles, *BM* and *CN* shall meet if produced on the same side of *DE* on which are the angles *MOP* and *NPO*.

PROPOSITION XVIII. THEOREM.

If one side of a triangle be greater than a second, the angle opposite the first must be greater than that opposite the second.

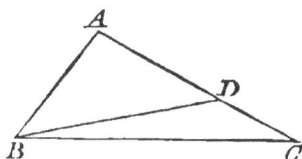

In △ *ABC*, let side *AC* be greater than *AB*.

Then must ∠ ABC be greater than ∠ ACB.

From *AC* cut off *AD=AB*, and join *BD*.　　　I. 3.

Then　　　　　　∵ *AB=AD*,

　　　∴ ∠ *ADB=* ∠ *ABD*,　　　I. A.

And ∵ *CD*, a side of △ *BDC*, is produced to *A*.

　　∴ ∠ *ADB* is greater than ∠ *ACB* ;　　I. 16

　∴ also ∠ *ABD* is greater than ∠ *ACB*.

Much more is ∠ *ABC* greater than ∠ *ACB*.

Q. E. D.

Ex. Shew that if two angles of a triangle be equal, the sides which subtend them are equal also (Eucl. I. 6).

PROPOSITION XIX. THEOREM.

If one angle of a triangle be greater than a second, the side opposite the first must be greater than that opposite the second.

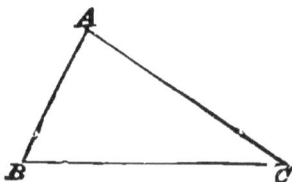

In △ *ABC*, let ∠ *ABC* be greater than ∠ *ACB*.

Then must AC be greater than AB.

For if *AC* be not greater than *AB*,

AC must either = *AB*, or be less than *AB*.

Now *AC* cannot = *AB*, for then   I. A.

∠ *ABC* would = ∠ *ACB*, which is not the case.

And *AC* cannot be less than *AB*, for then   I. 18.

∠ *ABC* would be less than ∠ *ACB*, which is not the case ;

∴ *AC* is greater than *AB*.

Q. E. D.

Ex. 1. In an obtuse-angled triangle, the greatest side is opposite the obtuse angle.

Ex. 2. *BC*, the base of an isosceles triangle *BAC*, is produced to any point *D* ; shew that *AD* is greater than *AB*.

Ex. 3. The perpendicular is the shortest straight line, which can be drawn from a given point to a given straight line ; and of others, that which is nearer to the perpendicular is less than one more remote.

Any two sides of a triangle are together greater than the third side.

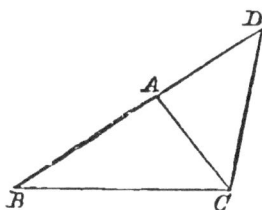

Let ABC be a $\triangle$.

Then any two of its sides must be together greater than the third side.

Produce BA to D, making $AD = AC$, and join DC.

Then $\because AD = AC$,

$\therefore \angle ACD = \angle ADC$, that is, $\angle BDC$.   I. A.

Now $\angle BCD$ is greater than $\angle ACD$;

$\therefore \angle BCD$ is also greater than $\angle BDC$;

$\therefore BD$ is greater than BC.   I. 19.

But $BD = BA$ and AD together ;

that is, $BD = BA$ and AC together ;

$\therefore BA$ and AC together are greater than BC.

Similarly it may be shewn that

AB and BC together are greater than AC,

and BC and CA .... AB.

             Q. E. D.

Ex. 1. Prove that any three sides of a quadrilateral figure are together greater than the fourth side.

Ex. 2. Shew that any side of a triangle is greater than the difference between the other two sides.

Ex. 3. Prove that the sum of the distances of any point from the angular points of a quadrilateral is greater than half the perimeter of the quadrilateral.

Ex. 4. If one side of a triangle be bisected, the sum of the two other sides shall be more than double of the line joining the vertex and the point of bisection.

PROPOSITION XXI. THEOREM.

If, from the ends of the side of a triangle, there be drawn two straight lines to a point within the triangle; these will be together less than the other sides of the triangle, but will contain a greater angle.

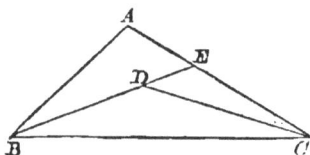

Let ABC be a $\triangle$, and from D, a pt. in the $\triangle$, draw st. lines to B and C.

Then will BD, DC together be less than BA, AC,
but $\angle BDC$ will be greater than $\angle BAC$.

Produce BD to meet AC in E.

Then BA, AE are together greater than BE.   I. 20.

Add to each EC.

Then BA, AC are together greater than BE, EC.

Again, DE, EC are together greater than DC.   I. 20

Add to each BD.

Then BE, EC are together greater than BD, DC.

And it has been shewn that BA, AC are together greater than BE, EC;

$\therefore BA$, AC are together greater than BD, DC.

Next, $\because \angle BDC$ is greater than $\angle DEC$,   I. 16.

and $\angle DEC$ is greater than $\angle BAC$,   I. 16.

$\therefore \angle BDC$ is greater than $\angle BAC$.

Q. E. D.

Ex. 1. Upon the base AB of a triangle ABC is described a quadrilateral figure $ADEB$, which is entirely within the triangle. Shew that the sides AC, CB of the triangle are together greater than the sides AD, DE, EB of the quadrilateral.

Ex. 2. Shew that the sum of the straight lines, joining the angles of a triangle with a point within the triangle, is less than the perimeter of the triangle, and greater than half the perimeter.

PROPOSITION XXII. PROBLEM.

To make a triangle, of which the sides shall be equal to three given straight lines, any two of which are together greater than the third.

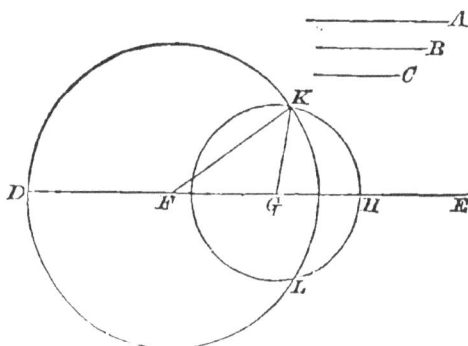

Let A, B, C be the three given lines, any two of which are together greater than the third.

It is required to make a $\triangle$ having its sides $= A$, B, C respectively.

Take a st. line DE of unlimited length.

In DE make $DF=A$, $FG=B$, and $GH=C$.   I. 3.

With centre F and distance FD, describe $\odot DKL$.

With centre G and distance GH, describe $\odot HKL$.

Join FK and GK.

Then $\triangle KFG$ has its sides $= A$, B, C respectively.

For $FK=FD$;   Def. 13.

$\therefore FK=A$;

and $GK=GH$;   Def. 13.

$\therefore GK=C$;

and $FG=B$;

$\therefore$ a $\triangle KFG$ has been described as reqd.   Q. E. F.

Ex. Draw an isosceles triangle having each of the equal sides double of the base.

PROPOSITION XXIII.   PROBLEM.

At a given point in a given straight line, to make an angle equal to a given angle.

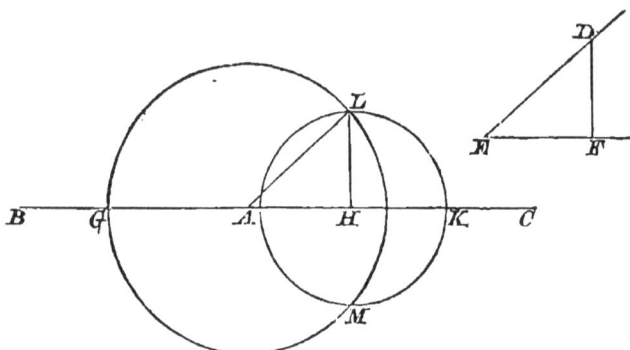

Let A be the given pt., BC the given line, DEF the given $\angle$.

It is reqd. to make at pt. A an angle $= \angle DEF$.

In ED, EF take any pts. D. F ; and join DF.

In AB, produced if necessary, make $AG = DE$.

In AC, produced if necessary, make $AH = EF$.

In HC, produced if necessary, make $HK = FD$.

With centre A, and distance AG, describe $\odot$ GLM.

With centre H, and distance HK, describe $\odot$ LKM.

Join AL and HL.

Then $\because$ $LA = AG$, $\therefore$ $LA = DE$;   Ax. 1.

and $\because$ $HL = HK$, $\therefore$ $HL = FD$.   Ax. 1.

Then in $\triangle$s LAH, DEF,

$\because$ $LA = DE$, and $AH = EF$, and $HL = FD$;

$\therefore$ $\angle LAH = \angle DEF$.   I. c.

$\therefore$ an angle LAH has been made at pt. A as was reqd.

Q. E. F.

NOTE.—We here give the proof of a theorem, necessary to the proof of Prop. XXIV. and applicable to several propositions in Book III.

PROPOSITION D. THEOREM.

Every straight line, drawn from the vertex of a triangle to the base, is less than the greater of the two sides, or than either, if they be equal.

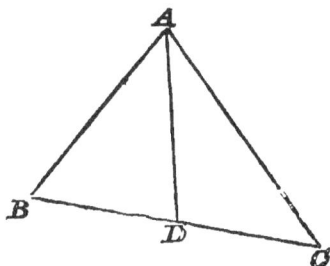

In the △ *ABC*, let the side *AC* be not less than *AB*.

Take any pt. *D* in *BC*, and join *AD*.

Then must *AD* be less than *AC*.

For ∵ *AC* is not less than *AB* ;

∴ ∠ *ABD* is not less than ∠ *ACD*.   I. A. and 18.

But ∠ *ADC* is greater than ∠ *ABD* ;   I. 16.

∴ ∠ *ADC* is greater than ∠ *ACD* ;

∴ *AC* is greater than *AD*.   I. 19.

Q. E. D.

Proposition XXIV. Theorem.

If two triangles have two sides of the one equal to two sides of the other, each to each, but the angle contained by the two sides of one of them greater than the angle contained by the two sides equal to them of the other ; the base of that which has the greater angle must be greater than the base of the other.

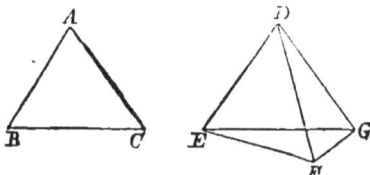

In the △s ABC, DEF,
let $AB = DE$ and $AC = DF$,
and let ∠ BAC be greater than ∠ EDF.
Then must BC be greater than EF.

Of the two sides DE, DF let DE be not greater than DF.*
At pt. D in st. line ED make ∠ EDG = ∠ BAC,   I. 23.
and make $DG = AC$ or DF, and join EG, GF.

Then ∵ $AB = DE$, and $AC = DG$, and ∠ BAC = ∠ EDG,
∴ $BC = EG$,   I. 4.

Again,   ∵ $DG = DF$,
∴ ∠ DFG = ∠ DGF ;   I. A.
∴ ∠ EFG is greater than ∠ DGF ;
much more then ∠ EFG is greater than ∠ EGF ;
∴ EG is greater than EF.   I. 19.
But $EG = BC$;
∴ BC is greater than EF.

Q. E. D.

---

* This line was added by Simson to obviate a defect in Euclid's proof. *Without* this condition, three distinct cases must be discussed. *With* the condition, we can prove that F must lie below EG.

For since DF is not less than DE, and DG is drawn equal to DF, DG is not less than DE.

Hence by Prop. D, any line drawn from D to meet EG is less than DG, and therefore DF, being equal to DG, must extend beyond EG.

For another method of proving the Proposition, see p. 115.

PROPOSITION XXV. THEOREM.

*If two triangles have two sides of the one equal to two sides
of the other, each to each, but the base of the one greater than
the base of the other ; the angle also, contained by the sides of
that which has the greater base, must be greater than the angle
contained by the sides equal to them of the other.*

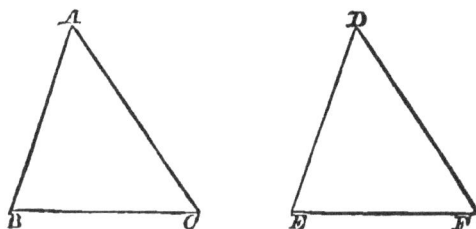

In the △s *ABC, DEF*,

let *AB=DE* and *AC=DF*,

and let *BC* be greater than *EF*.

Then must ∠ BAC be greater than ∠ EDF.

For ∠ *BAC* is greater than, equal to, or less than ∠ *EDF*.

Now ∠ *BAC* cannot= ∠ *EDF*,

for then, by I. 4, *BC* would=*EF* ; which is not the case.

And ∠ *BAC* cannot be less than ∠ *EDF*,

for then, by I. 24, *BC* would be less than *EF* ; which is
not the case ;

∴ ∠ *BAC* must be greater than ∠ *EDF*.

Q. E. D.

NOTE.—In Prop. XXVI. Euclid includes two cases, in which
two triangles are equal in all respects ; viz., when the following
parts are equal in the two triangles :

1. Two angles and the side between them.

2. Two angles and the side opposite one of them.

Of these we have already proved the first case, in **Prop. B**,
so that we have only the second case left, to form the subject
of Prop. XXVI., which we shall **prove by the** method of
superposition.

For Euclid's proof of Prop. XXVI , see p. 114-115.

Proposition XXVI. Theorem.

If two triangles have two angles of the one equal to two angles of the other, each to each, and one side equal to one side, those sides being opposite to equal angles in each; then must the triangles be equal in all respects.

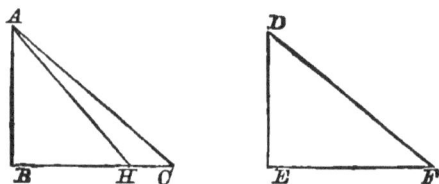

In △s ABC, DEF,

let $\angle ABC = \angle DEF$, and $\angle ACB = \angle DFE$, and $AB = DE$.

Then must $BC = EF$, and $AC = DF$, and $\angle BAC = \angle EDF$.

Suppose △DEF to be applied to △ABC,

so that D coincides with A, and DE falls on AB.

Then ∵ $DE = AB$, ∴ E will coincide with B ;

and ∵ $\angle DEF = \angle ABC$, ∴ EF will fall on BC.

Then must F coincide with C: for, if not,

let F fall *between* B and C, at the pt. H.   Join AH.

Then   ∵ $\angle AHB = \angle DFE$,   I. 4.

∴ $\angle AHB = \angle ACB$,

the extr. $\angle$ = the intr. and opposite $\angle$, which is impossible.

∴ F does not fall between B and C.

Similarly, it may be shewn that F does not fall on BC produced.

∴ F coincides with C, and ∴ $BC = EF$;

∴ $AC = DF$, and $\angle BAC = \angle EDF$,   I. 4

and ∴ the triangles are equal in all respects.

Q. E. D.

Miscellaneous Exercises on Props. I. to XXVI.

1. M is the middle point of the base BC of an isosceles triangle ABC, and N is a point in AC. Shew that the difference between MB and MN is less than that between AB and AN.

2. ABC is a triangle, and the angle at A is bisected by a straight line which meets BC at D; shew that BA is greater than BD, and CA greater than CD.

3. AB, AC are straight lines meeting in A, and D is a given point. Draw through D a straight line cutting off equal parts from AB, AC.

4. Draw a straight line through a given point, to make equal angles with two given straight lines which meet.

5. A given angle BAC is bisected; if CA be produced to G and the angle BAG bisected, the two bisecting lines are at right angles.

6. Two straight lines are drawn to the base of a triangle from the vertex, one bisecting the vertical angle, and the other bisecting the base. Prove that the latter is the greater of the two lines.

7. Shew that Prop. XVII. may be proved without producing a side of the triangle.

8. Shew that Prop. XVIII. may be proved by means of the following construction : cut off $AD = AB$, draw AE, bisecting $\angle BAC$ and meeting BC in E, and join DE.

9. Shew that Prop. XX. can be proved, without producing one of the sides of the triangle, by bisecting one of the angles.

10. Given two angles of a triangle and the side adjacent to them, construct the triangle.

11. Shew that the perpendiculars, let fall on two sides of a triangle from any point in the straight line bisecting the angle contained by the two sides, are equal.

We conclude Section I. with the proof (omitted by Euclid) of another case in which two triangles are equal in all respects.

PROPOSITION E. THEOREM.

If two triangles have one angle of the one equal to one angle of the other, and the sides about a second angle in each equal: then, if the third angles in each be both acute, both obtuse, or if one of them be a right angle, the triangles are equal in all respects.

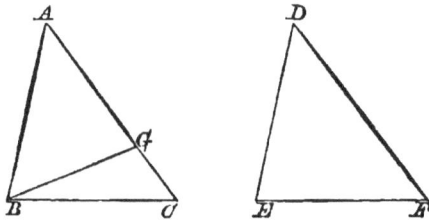

In the △s ABC, DEF, let $\angle BAC = \angle EDF$, $AB = DE$, $BC = EF$, and let $\angle$ s ACB, DFE be both acute, both obtuse, or let one of them be a right angle.

Then must △s ABC, DEF be equal in all respects.

For if AC be not $= DF$, make $AG = DF$; and join BG.
Then in △s BAG, EDF,
∵ $BA = ED$, and $AG = DF$, and $\angle BAG = \angle EDF$,
∴ $BG = EF$ and $\angle AGB = \angle DFE$.   I. 4.
But $BC = EF$, and ∴ $BG = BC$;
∴ $\angle BCG = \angle BGC$ I. A.
First, let $\angle ACB$ and $\angle DFE$ be both acute,
    then $\angle AGB$ is acute, and ∴ $\angle BGC$ is obtuse ;   I. 13.
∴ $\angle BCG$ is obtuse, which is contrary to the hypothesis.
Next, let $\angle ACB$ and $\angle DFE$ be both obtuse,
    then $\angle AGB$ is obtuse, and ∴ $\angle BGC$ is acute ;   I. 13.
∴ $\angle BCG$ is acute, which is contrary to the hypothesis.

Lastly, let one of the third angles ACB, DFE be a right angle.

If $\angle ACB$ be a rt. $\angle$,

　　　then $\angle BGC$ is also a rt. $\angle$;　　　　　　　I. A.

$\therefore \angle$ s BCG, BGC together $=$ two rt. $\angle$ s, which is impossible.　　　　　　　　　　　　　　　　　I. 17.

Again, if $\angle DFE$ be a rt. $\angle$,

　　　then $\angle AGB$ is a rt. $\angle$, and $\therefore \angle BGC$ is a rt. $\angle$.　　I. 13.

Hence $\angle BCG$ is also a rt. $\angle$.

$\therefore \angle$ s BCG, BGC together $=$ two rt. $\angle$ s, which is impossible.

I. 17.

Hence AC is equal to DF,

and the $\triangle$ s ABC, DEF are equal in all respects.

Q. E. D.

COR. From the first case of this proposition we deduce the following important theorem :

If two right-angled triangles have the hypotenuse and one side of the one equal respectively to the hypotenuse and one side of the other, the triangles are equal in all respects.

NOTE. In the enunciation of Prop. E, if, instead of the words *if one of them be a right angle*, we put the words *both right angles,* this case of the proposition would be identical with I. 26.

SECTION II.

The Theory of Parallel Lines.

INTRODUCTION.

WE have detached the Propositions, in which Euclid treats of Parallel Lines, from those which precede and follow them in the First Book, in order that the student may have a clearer notion of the difficulties attending this division of the subject, and of the way in which Euclid proposes to meet them.

We must first explain some technical terms used in this Section.

If a straight line *EF* cut two other straight lines *AB*, *CD*, it makes with those lines eight angles, to which particular names are given.

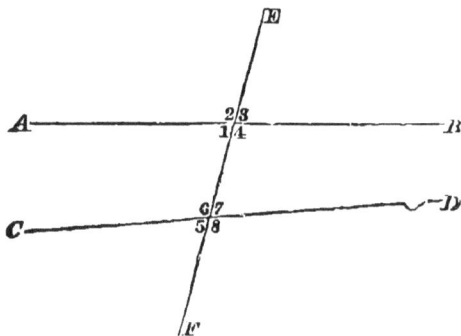

The angles numbered 1, 4, 6, 7 are called *Interior* angles.
.... 2, 3, 5, 8 ...... *Exterior* ....

The angles marked 1 and 7 are called *alternate* angles.

The angles marked 4 and 6 are also called alternate angles.

The pairs of angles 1 and 5, 2 and 6, 4 and 8, 3 and 7 are called *corresponding* angles.

NOTE. From I. 13 it is clear that the angles 1, 4, 6, 7 are together equal to four right angles.

PROPOSITION XXVII. THEOREM.

If a straight line, falling upon two other straight lines, make the alternate angles equal to one another; these two straight lines must be parallel.

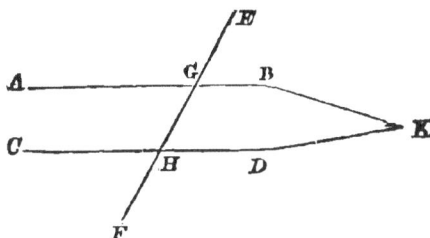

Let the st. line *EF*, falling on the st. lines *AB*, *CD*,

make the alternate ∠ s *AGH*, *GHD* equal.

Then must AB be ∥ to CD.

For if not, *AB* and *CD* will meet, if produced, either towards *B*, *D*, or towards *A*, *C*.

Let them be produced and meet towards *B*, *D* in *K*.

Then *GHK* is a △ ;

and ∴ ∠ *AGH* is greater than ∠ *GHD*.   I. 16.

But   ∠ *AGH* = ∠ *GHD*,   Hyp.

which is impossible.

∴ *AB*, *CD* do not meet when produced towards *B*, *D*.

In like manner it may be shewn that they do not meet when produced towards *A*. *C*.

∴ *AB* and *CD* are parallel.   Def. 26.

Q. E. D.

Proposition XXVIII. Theorem.

If a straight line, falling upon two other straight lines, make the exterior angle equal to the interior and opposite upon the same side of the line, or make the interior angles upon the same side together equal to two right angles; the two straight lines are parallel to one another.

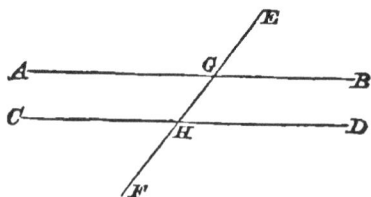

Let the st. line EF, falling on st. lines AB, CD, make

     I. $\angle EGB$ = corresponding $\angle GHD$, or

     II. $\angle$ s BGH, GHD together = two rt. $\angle$ s.

Then, in either case, AB must be $\parallel$ to CD.

I.

     $\because \angle EGB$ is given = $\angle GHD$,   Hyp.

and $\angle EGB$ is known to be = $\angle AGH$,   I. 15.

     $\therefore \angle AGH = \angle GHD$;

and these are alternate $\angle$ s ;

     $\therefore AB$ is $\parallel$ to CD.   I. 27.

II. $\because \angle$ s BGH, GHD together = two rt. $\angle$ s,   Hyp.

and $\angle$ s BGH, AGH together = two rt. $\angle$ s,   I. 13.

$\therefore \angle$ s BGH, AGH together = $\angle$ s BGH, GHD together ;

     $\therefore \angle AGH = \angle GHD$;

     $\therefore AB$ is $\parallel$ to CD.   I. 27.

Q. E. D.

In the place of Euclid's Sixth Postulate many modern writers on Geometry propose, as more evident to the senses, the following Postulate :—

" *Two straight lines which cut one another cannot* BOTH *be parallel to the same straight line.*"

If this be assumed, we can prove Post. 6, as a Theorem, thus :

Let the line EF falling on the lines AB, CD make the $\angle$ s BGH, GHD together less than two rt. $\angle$ s. Then must AB, CD meet when produced towards B, D.

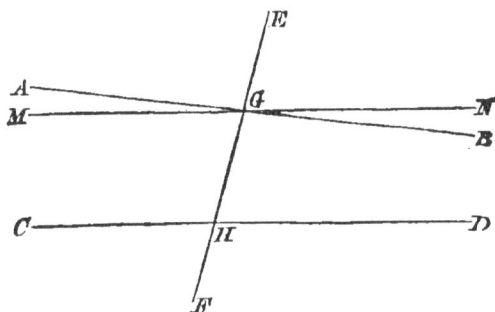

For if not, suppose AB and CD to be parallel.

Then ∵ $\angle$ s AGH, BGH together $=$ two rt. $\angle$ s,   I. 13.

and $\angle$ s GHD, BGH are together less than two rt. $\angle$ s,

∴ $\angle AGH$ is greater than $\angle GHD$.

Make $\angle MGH = \angle GHD$, and produce MG to N.

Then ∵ the alternate $\angle$ s MGH, GHD are equal,

∴ MN is ∥ to CD.   I. 27.

Thus two lines MN, B which cut one another are both parallel to CD, which is impossible.

∴ AB and CD are not parallel.

It is also clear that they meet towards B, D, because GB lies between GN and HD.

Q. E. D

Proposition XXIX. Theorem.

If a straight line fall upon two parallel straight lines, it makes the two interior angles upon the same side together equal to two right angles, and also the alternate angles equal to one another, and also the exterior angle equal to the interior and opposite upon the same side.

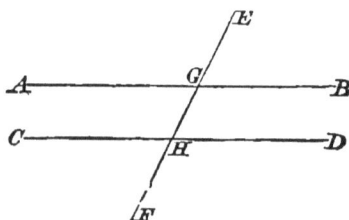

Let the st. line EF fall on the parallel st. lines AB, CD.

Then must

    I. ∠s BGH, GHD together = two rt. ∠s.

    II. ∠ AGH = alternate ∠ GHD.

    III. ∠ EGB = corresponding ∠ GHD.

I. ∠s BGH, GHD cannot be together *less* than two rt. ∠s,

    for then AB and CD would meet if produced towards B and D,   Post. 6.

    which cannot be, for they are parallel.

  Nor can ∠s BGH, GHD be together *greater* than two rt. ∠s,

    for then ∠s AGH, GHC would be together less than two rt. ∠s,   I. 13.

    and AB, CD would meet if produced towards A and C Post. 6

    which cannot be, for they are parallel,

  ∴ ∠s BGH, GHD together = two rt. ∠s.

II. ∵ ∠s BGH, GHD together = two rt. ∠s,

    and ∠s BGH, AGH together = two rt. ∠s,   I. 13.

  ∴ ∠s BGH, AGH together = ∠s BGH, GHD together,

    and ∴ ∠ AGH = ∠ GHD.   Ax. 3.

III. ∵ ∠ AGH = ∠ GHD,

    and ∠ AGH = ∠ EGB,   I. 15.

  ∴ ∠ EGB = ∠ GHD.   Ax. 1

Q. E. D.

1. If through a point, equidistant from two parallel straight lines, two straight lines be drawn cutting the parallel straight lines ; they will intercept equal portions of the parallel lines.

2. If a straight line be drawn, bisecting one of the angles of a triangle, to meet the opposite side ; the straight lines drawn from the point of section, parallel to the other sides and terminated by those sides, will be equal.

3. If any straight line joining two parallel straight lines be bisected, any other straight line, drawn through the point of bisection to meet the two lines, will be bisected in that point.

NOTE. One Theorem (A) is said to be the *converse* of another Theorem (B), when the hypothesis in (A) is the conclusion in (B), and the conclusion in (A) is the hypothesis in (B).

For example, the Theorem I. A. may be stated thus :

Hypothesis. If two sides of a triangle be equal.

Conclusion. The angles opposite those sides must also be equal.

The converse of this is the Theorem I. B. Cor. :

Hypothesis. If two angles of a triangle be equal.

Conclusion. The sides opposite those angles must also be equal.

The following are other instances :

Postulate VI. is the converse of I. 17.

I. 29 is the converse of I. 27 and 28.

Proposition XXX. Theorem.

Straight lines which are parallel to the same straight line are parallel to one another.

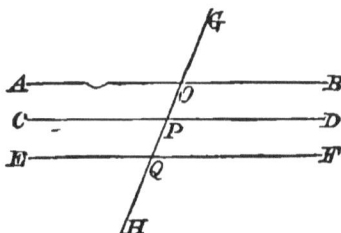

Let the st. lines AB, CD be each ∥ to EF.

Then must AB be ∥ to CD.

Draw the st. line GH, cutting AB, CD, EF in the pts. O, P, Q.

Then ∵ GH cuts the ∥ lines AB, EF,

∴ ∠ AOP = alternate ∠ PQF.   I. 29.

And ∵ GH cuts the ∥ lines CD, EF,

∴ extr. ∠ OPD = intr. ∠ PQF ;   I. 29.

∴ ∠ AOP = ∠ OPD ;

and these are alternate angles ;

∴ AB is ∥ to CD.   I. 27.

Q. E. D.

The following Theorems are important. They admit of easy proof, and are therefore left as Exercises for the student.

1. If two straight lines be parallel to two other straight lines, each to each, the first pair make the same angles with one another as the second.

2. If two straight lines be perpendicular to two other straight lines, each to each, the first pair make the same angles with one another as the second.

To draw a straight line through a given point parallel to a given straight line.

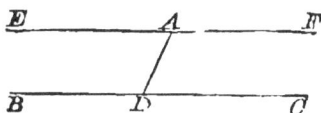

Let A be the given pt. and BC the given st. line.

It is required to draw through A a st. line $\parallel$ to BC.

In BC take any pt. D, and join AD.

$\qquad$ Make $\angle DAE = \angle ADC$. $\qquad\qquad$ I. 23.

Produce EA to F. Then EF shall be $\parallel$ to BC.

For $\because$ AD, meeting EF and BC, makes the alternate angles equal, that is, $\angle EAD = \angle ADC$,

$\qquad\qquad\qquad \therefore EF$ is $\parallel$ to BC. $\qquad\qquad$ I. 27

$\therefore$ a st. line has been drawn through $A \parallel$ to BC.

$\qquad\qquad\qquad\qquad\qquad\qquad$ Q. E. F.

Ex. 1. From a given point draw a straight line, to make an angle with a given straight line that shall be equal to a given angle.

Ex. 2. Through a given point A draw a straight line ABC, meeting two parallel straight lines in B and C, so that BC may be equal to a given straight line.

Proposition XXXII. Theorem.

If a side of any triangle be produced, the exterior angle is equal to the two interior and opposite angles, and the three interior angles of every triangle are together equal to two right angles.

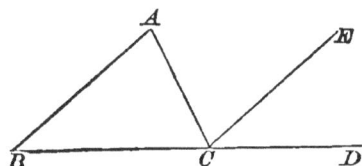

Let ABC be a △, and let one of its sides, BC, be produced to D.

Then will

  I. $\angle ACD = \angle$ s ABC, BAC *together*.

  II. $\angle$ s ABC, BAC, ACB *together* = two rt. $\angle$ s.

         From C draw $CE \parallel$ to AB.   I. 31.

Then I. ∵ BD meets the ∥s EC, AB,

       ∴ extr. $\angle ECD$ = intr. $\angle ABC$.   I. 29.

And ∵ AC meets the ∥s EC, AB,

       ∴ $\angle ACE$ = alternate $\angle BAC$.   I. 29.

∴ $\angle$ s ECD, ACE together = $\angle$ s ABC, BAC together ;

       ∴ $\angle ACD = \angle$ s ABC, BAC together.

And II. ∵ $\angle$ s ABC, BAC together = $\angle ACD$,

      to each of these equals add $\angle ACB$;

then $\angle$ s ABC, BAC, ACB together = $\angle$ s ACD, ACB together,

       ∴ $\angle$ s ABC, BAC, ACB together = two rt. $\angle$ s.   I. 13.

                               Q. E. D.

Ex. 1. In an acute-angled triangle, any two angles are greater than the third.

Ex. 2. The straight line, which bisects the external vertical angle of an isosceles triangle is parallel to the base.

Ex. 3. If the side *BC* of the triangle *ABC* be produced to *D*, and *AE* be drawn bisecting the angle *BAC* and meeting *BC* in *E* ; shew that the angles *ABD*, *ACD* are together double of the angle *AED*.

Ex. 4. If the straight lines bisecting the angles at the base of an isosceles triangle be produced to meet ; shew that they will contain an angle equal to an exterior angle at the base of the triangle.

Ex. 5. If the straight line bisecting the external angle of a triangle be parallel to the base ; prove that the triangle is isosceles.

The following Corollaries to Prop. 32 were first given in Simson's Edition of Euclid.

Cor. 1. *The sum of the interior angles of any rectilinear figure together with four right angles is equal to twice as many right angles as the figure has sides.*

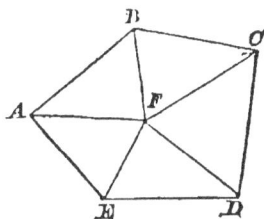

Let *ABCDE* be any rectilinear figure.

Take any pt. *F* within the figure, and from *F* draw the st. lines *FA, FB, FC, FD, FE* to the angular pts. of the figure

Then there are formed as many ∠ s as the figure has sides.

The three ∠ s in *each* of these △ s together = two rt. ∠ s.

∴ *all* the ∠ s in these △ s together = twice as many right ∠ s as there are △ s, that is, twice as many right ∠ s as the figure has sides.

Now angles of all the △ s = ∠ s at *A, B, C, D, E* and ∠ s at *F*,

that is,   = ∠ s of the figure and ∠ s at *F*,

and ∴   = ∠ s of the figure and four rt. ∠ s.   I. 15. Cor. 2.

∴ ∠ s of the figure and four rt. ∠ s = twice as many rt. ∠ s as the figure has sides.

Cor. 2. *The exterior angles of any convex rectilinear figure, made by producing each of its sides in succession, are together equal to four right angles.*

Every interior angle, as ABC, and its adjacent exterior angle, as ABD, together are = two rt. ∠ s.

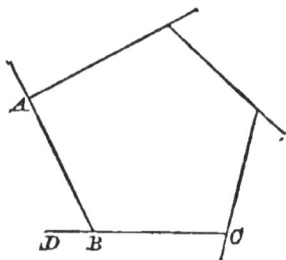

∴ all the intr. ∠ s together with all the extr. ∠ s
= twice as many rt. ∠ s as the figure has sides.

But all the intr. ∠ s together with four rt. ∠ s
= twice as many rt. ∠ s as the figure has sides.

∴ all the intr. ∠ s together with all the extr. ∠ s
= all the intr. ∠ s together with four rt. ∠ s.

∴ all the extr. ∠ s = four rt. ∠ s.

NOTE. The latter of these corollaries refers only to *convex* figures, that is, figures in which every interior angle is less than two right angles. When a figure contains an angle greater

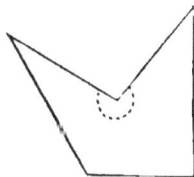

than two right angles, as the angle marked by the dotted line in the diagram, this is called a reflex *angle*. See p. 149.

Ex. 1. The exterior angles of a quadrilateral made by producing the sides successively are together equal to the interior angles.

Ex. 2. Prove that the interior angles of a hexagon are equal to eight right angles.

Ex. 3. Shew that the angle of an equiangular pentagon is $\frac{6}{5}$ of a right angle.

Ex. 4. How many sides has the rectilinear figure, the sum of whose interior angles is double that of its exterior angles ?

Ex. 5. How many sides has an equiangular polygon, four of whose angles are together equal to seven right angles ?

PROPOSITION XXXIII. THEOREM.

The straight lines which join the extremities of two equal and parallel straight lines, towards the same parts, are also them-selves equal and parallel.

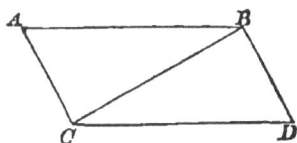

Let the equal and ∥ st. lines AB, CD be joined towards the same parts by the st. lines AC, BD.

Then must AC and BD be equal and ∥.

      Join BC.

Then   ∵ AB is ∥ to CD,

      ∴ ∠ABC=alternate ∠DCB.   I. 29.

Then in △s ABC, BCD,

∵ AB=CD, and BC is common, and ∠ABC= ∠DCB,

      ∴ AC=BD, and ∠ACB= ∠DBC.   I. 4.

Then ∵ BC, meeting AC and BD,

      makes the alternate ∠s ACB, DBC equal,

      ∴ AC is ∥ to BD.

                          Q. E. D.

Miscellaneous Exercises on Sections I. and II.

1. If two exterior angles of a triangle be bisected by straight lines which meet in O; prove that the perpendiculars from O on the sides, or the sides produced, of the triangle are equal.

2. Trisect a right angle.

3. The bisectors of the three angles of a triangle meet in one point.

4. The perpendiculars to the three sides of a triangle drawn from the middle points of the sides meet in one point.

5. The angle between the bisector of the angle BAC of the triangle ABC and the perpendicular from A on BC, is equal to half the difference between the angles at B and C.

6. If the straight line AD bisect the angle at A of the triangle ABC, and BDE be drawn perpendicular to AD, and meeting AC, or AC produced, in E; shew that BD is equal to DE.

7. Divide a right-angled triangle into two isosceles triangles.

8. AB, CD are two given straight lines. Through a point E between them draw a straight line GEH, such that the intercepted portion GH shall be bisected in E.

9. The vertical angle O of a triangle OPQ is a right, acute, or obtuse angle, according as OR, the line bisecting PQ, is equal to, greater or less than the half of PQ.

10. Shew by means of Ex. 9 how to draw a perpendicular to a given straight line from its extremity without producing it.

SECTION III.

On the Equality of Rectilinear Figures in respect of Area.

THE amount of space enclosed by a Figure is called the Area of that figure.

Euclid calls two figures *equal* when they enclose the same amount of space. They may be dissimilar in shape, but if the areas contained within the boundaries of the figures be the same, then he calls the figures *equal*. He regards a triangle, for example, as a figure having sides and angles and area, and he proves in this section that two triangles may have equality of area, though the sides and angles of each may be unequal.

Coincidence of their boundaries is a test of the equality of all geometrical magnitudes, as we explained in Note 1, page 14.

In the case of lines and angles it is the only test : in the case of *figures* it is *a test, but not the only test ;* as we shall shew in this Section.

The sign =, standing between the symbols denoting two *figures*, must be read *is equal in area to.*

Before we proceed to prove the Propositions included in this Section, we must complete the list of Definitions required in Book I., continuing the numbers prefixed to the definitions in page 6.

DEFINITIONS.

XXVII. A PARALLELOGRAM is a four-sided figure whose opposite sides are parallel.

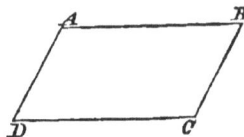

For brevity we often designate a parallelogram by two letters only, which mark opposite angles. Thus we call the figure in the margin the parallelogram *A C*.

XXVIII. A Rectangle is a parallelogram, having one of its angles a right angle.

Hence by I. 29, *all* the angles of a rectangle are right angles.

XXIX. A RHOMBUS is a parallelogram, having its sides equal.

XXX. A SQUARE is a parallelogram, having its sides equal and one of its angles a right angle.

Hence, by I. 29, *all* the angles of a square are right angles.

XXXI. A TRAPEZIUM is a four-sided figure of which two sides only are parallel.

XXXII. A DIAGONAL of a four-sided figure is the straight line joining two of the opposite angular points.

XXXIII. The ALTITUDE of a Parallelogram is the perpendicular distance of one of its sides from the side opposite,
regarded as the Base.

The altitude of a triangle is the perpendicular distance of
one of its angular points from the side opposite, regarded as
the base.

Thus if *ABCD* be a parallelogram, and *AE* a perpendicular
let fall from *A* to *CD*, *AE* is the *altitude* of the parallelogram,
and also of the triangle *ACD*.

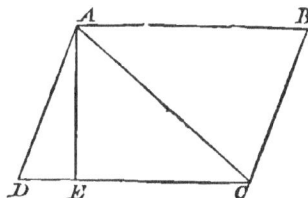

If a perpendicular be let fall from *B* to *DC* produced, meeting *DC* in *F*, *BF* is the altitude of the parallelogram.

EXERCISES.

Prove the following theorems :

1. The diagonals of a square make with each of the sides
an angle equal to half a right angle.

2. If two straight lines bisect each other, the lines joining
their extremities will form a parallelogram.

3. Straight lines bisecting two adjacent angles of a parallelogram intersect at right angles.

4. If the straight lines joining two opposite angular points
of a parallelogram bisect the angles, the parallelogram has all
its sides equal.

5. If the opposite angles of a quadrilateral be equal, the
quadrilateral is a parallelogram.

6. If two opposite sides of a quadrilateral figure be equal to
one another, and the two remaining sides be also equal to one
another, the figure is a parallelogram.

7. If one angle of a rhombus be equal to two-thirds of two
right angles, the diagonal drawn from that angular point
divides the rhombus into two equilateral triangles.

Proposition XXXIV. Theorem.

The opposite sides and angles of a parallelogram are equal to one another, and the diagonal bisects it.

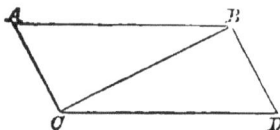

Let $ABDC$ be a $\square$, and BC a diagonal of the $\square$.

Then must $\quad AB=DC$ and $AC=DB$,

and $\quad \angle BAC = \angle CDB$, and $\angle ABD = \angle ACD$

and $\quad\quad\quad \triangle ABC = \triangle DCB$.

For ∵ AB is ∥ to CD, and BC meets them,

$\quad \therefore \angle ABC=$ alternate $\angle DCB$;   I. 29.

and ∵ AC is ∥ to BD, and BC meets them,

$\quad \therefore \angle ACB=$ alternate $\angle DBC$.   I. 29.

Then in $\triangle$ s ABC, DCB,

$\quad ∵ \angle ABC = \angle DCB$, and $\angle ACB = \angle DBC$,

and BC is common, a side adjacent to the equal $\angle$ s in each ;

$\quad \therefore AB=DC$, and $AC=DB$, and $\angle BAC = \angle CDB$,

and $\triangle ABC = \triangle DCB$.   I. B.

Also ∵ $\angle ABC = \angle DCB$, and $\angle DBC = \angle ACB$,

$\quad \therefore \angle$ s ABC, DBC together $= \angle$ s DCB, ACB together,

that is, $\quad\quad\quad \angle ABD = \angle ACD$,

Q. E. D.

Ex. 1. Shew that the diagonals of a parallelogram bisect each other.

Ex. 2. Shew that the diagonals of a rectangle are equal.

PROPOSITION XXXV. THEOREM.

Parallelograms on the same base and between the same parallels are equal.

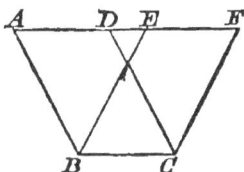

Let the ▱s $ABCD$, $EBCF$ be on the same base BC and between the same ‖s AF, BC.

Then must ▱ $ABCD =$ ▱ $EBCF$.

CASE I. If AD, EF have no point common to both,

Then in the △s FDC, EAB,

$\because$ extr. $\angle FDC =$ intr. $\angle EAB$, I. 29.

and intr. $\angle DFC =$ extr. $\angle AEB$, I. 29.

and $DC = AB$, I. 34.

$\therefore \triangle FDC = \triangle EAB$. I. 26.

Now ▱ $ABCD$ with △ $FDC =$ figure $ABCF$;

and ▱ $EBCF$ with △ $EAB =$ figure $ABCF$;

$\therefore$ ▱ $ABCD$ with △ $FDC =$ ▱ $EBCF$ with △ EAB :

$\therefore$ ▱ $ABCD =$ ▱ $EBCF$.

CASE II. If the sides AD, EF overlap one another

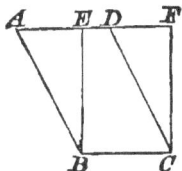

the same method of proof applies.

CASE III. If the sides opposite to BC be terminated in the same point D,

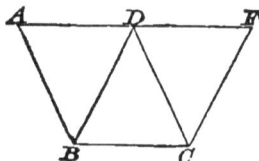

the same method of proof is applicable,
but it is easier to reason thus :

Each of the $\square$s is double of $\triangle BDC$;   I. 34.

$\therefore \square ABCD = \square DBCF.$

Q. E. D.

PROPOSITION XXXVI. THEOREM.

Parallelograms on equal bases, and between the same parallels, are equal to one another.

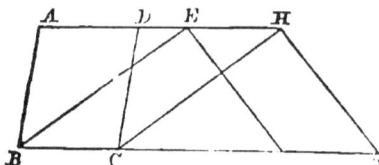

Let the $\square$s $ABCD$, $EFGH$ be on equal bases BC, FG, and between the same ∥s AH, BG.

Then must $\square ABCD = \square EFGH.$

Join BE, CH.

Then $\because BC = FG,$ Hyp.

and $EH = FG$;   I. 34.

$\therefore BC = EH$;

and BC is ∥ to EH.   Hyp.

$\therefore EB$ is ∥ to CH ;   I. 33.

$\therefore EBCH$ is a parallelogram.

Now $\square EBCH = \square ABCD$,   I. 35.

$\because$ they are on the same base BC and between the same ∥s ;

and $\square EBCH = \square EFGH$,   I. 35.

$\because$ they are on the same base EH and between the same ∥s ,

$\therefore \square ABCD = \square EFGH.$

Q. E. D.

PROPOSITION XXXVII. THEOREM.

Triangles upon the same base, and between the same parallels, are equal to one another.

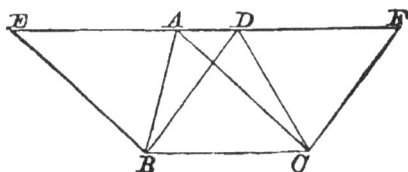

Let △s *ABC*, *DBC* be on the same base *BC* and between the same ‖s *AD*, *BC*.

Then must △ *ABC*= △ *DBC*.

From *B* draw *BE* ‖ to *CA* to meet *DA* produced in *E*.

From *C* draw *CF* ‖ to *BD* to meet *AD* produced in *F*.

Then *EBCA* and *FCBD* are parallelograms,

and ▱ *EBCA*=▱ *FCBD*,   I. 35.

∵ they are on the same base and between the same ‖s.

Now   △ *ABC* is half of ▱ *EBCA*,   I. 34.

and △ *DBC* is half of ▱ *FCBD*;   I. 34.

∴ △ *ABC*= △ *DBC*.   Ax. 7.

Q. E. D.

Ex. 1. If *P* be a point in a side *AB* of a parallelogram *ABCD*, and *PC*, *PD* be joined, the triangles *PAD*, *PBC* are together equal to the triangle *PDC*.

Ex. 2. If *A*, *B* be points in one, and *C*, *D* points in another of two parallel straight lines, and the lines *AD*, *BC* intersect in *E*, then the triangles *AEC*, *BED* are equal.

PROPOSITION XXXVIII. THEOREM.

Triangles upon equal bases, and between the same parallels, are equal to one another.

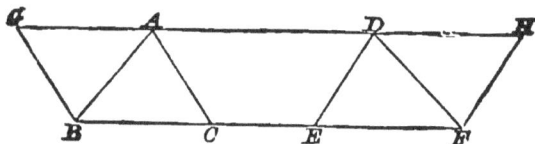

Let △s *ABC, DEF* be on equal bases, *BC, EF,* and between the same ‖s *BF, AD.*

Then must △ *ABC*= △ *DEF.*

From *B* draw *BG* ‖ to *CA* to meet *DA* produced in *G.*

From *F* draw *FH* ‖ to *ED* to meet *AD* produced in *H.*

Then *CG* and *EH* are parallelograms, and they are equal,

∵ they are on equal bases *BC, EF,* and between the same ‖s *BF, GH.*   I. 36

Now △ *ABC* is half of ▭ *CG,*

and △ *DEF* is half of ▭ *EH* ;

∴ △*ABC*= △*DEF.*   Ax. 7.

Q. E. D.

Ex. 1. Shew that a straight line, drawn from the vertex of a triangle to bisect the base, divides the triangle into two equal parts.

Ex. 2. In the equal sides *AB, AC* of an isosceles triangle *ABC* points *D, E* are taken such that *BD*=*AE.* Shew that the triangles *CBD, ABE* are equal.

PROPOSITION XXXIX. THEOREM.

Equal triangles upon the same base, and upon the same side of it, are between the same parallels.

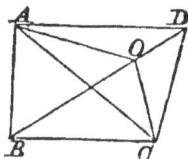

Let the equal △s *ABC*, *DBC* be on the same base *BC*, and on the same side of it.

Join *AD*.

Then must AD be ∥ to BC.

For if not, through *A* draw *AO* ∥ to *BC*, so as to meet *BD*, or *BD* produced, in *O*, and join *OC*.

Then ∵ △s *ABC*, *OBC* are on the same base and between the same ∥s,

$$∴ △ ABC = △ OBC. \qquad\qquad \text{I. 37.}$$

But $\qquad\qquad △ ABC = △ DBC; \qquad\qquad$ Hyp.

$$∴ △ OBC = △ DBC,$$

the less = the greater, which is impossible ;

$$∴ AO \text{ is not } ∥ \text{ to } BC.$$

In the same way it may be shewn that no other line passing through *A* but *AD* is ∥ to *BC* ;

$$∴ AD \text{ is } ∥ \text{ to } BC.$$

Q. E. D.

Ex. 1. *AD* is parallel to *BC* ; *AC*, *BD* meet in *E* ; *BC* is produced to *P* so that the triangle *PEB* is equal to the triangle *ABC* : shew that *PD* is parallel to *AC*.

Ex. 2. If of the four triangles into which the diagonals divide a quadrilateral, two opposite ones are equal, the quadrilateral has two opposite sides parallel.

S. E.   5

PROPOSITION XL. THEOREM.

Equal triangles upon equal bases, in the same straight line,
and towards the same parts, are between the same parallels.

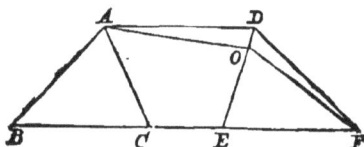

Let the equal △s *ABC*, *DEF* be on equal bases *BC*, *EF*
in the same st. line *BF* and towards the same parts.

Join *AD*.

Then must AD be ∥ to BF.

For if not, through *A* draw *AO* ∥ to *BF*, so as to meet *ED*.
or *ED* produced, in *O*, and join *OF*.

Then △ *ABC*= △ *OEF*, ∵ they are on equal bases and
between the same ∥s.   I. 38.

But   △ *ABC*= △ *DEF* ;   Hyp.

∴ △ *OEF*= △ *DEF*,

the less = the greater, which is impossible.

∴ *AO* is not ∥ to *BF*.

In the same way it may be shewn that no other line passing
through *A* but *AD* is ∥ to *BF*,

∴ *AD* is ∥ to *BF*.

Q. E. D.

Ex. 1. The straight line, joining the points of bisection of
two sides of a triangle, is parallel to the base, and is equal to
half the base.

Ex. 2. The straight lines, joining the middle points of the
sides of a triangle, divide it into four equal triangles.

If a parallelogram and a triangle be upon the same base, and between the same parallels, the parallelogram is double of the triangle.

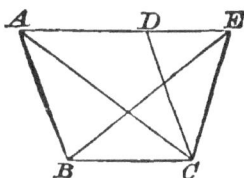

Let the $\square$ $ABCD$ and the $\triangle EBC$ be on the same base BC and between the same ||s AE, BC.

Then must $\square$ $ABCD$ be double of $\triangle EBC$.

Join AC.

Then $\triangle ABC = \triangle EBC$, $\because$ they are on the same base and between the same ||s ;   I. 37.

and $\square$ $ABCD$ is double of $\triangle ABC$, $\because$ AC is a diagonal of $ABCD$;   I. 34.

$\therefore$ $\square$ $ABCD$ is double of $\triangle EBC$.

Q. E. D.

Ex. 1. If from a point, without a parallelogram, there be drawn two straight lines to the extremities of the two opposite sides, between which, when produced, the point does not lie, the difference of the triangles thus formed is equal to half the parallelogram.

Ex. 2. The two triangles, formed by drawing straight lines from any point within a parallelogram to the extremities of its opposite sides, are together half of the parallelogram.

PROPOSITION XLII.   PROBLEM.

To describe a parallelogram that shall be equal to a given triangle, and have one of its angles equal to a given angle.

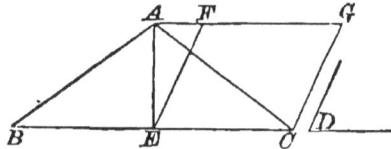

Let ABC be the given $\triangle$, and D the given $\angle$.

It is required to describe a $\square$ equal to $\triangle ABC$, having one of its $\angle s = \angle D$.

Bisect BC in E and join AE.	I. 10.
At E make $\angle CEF = \angle D$.	I. 23.

Draw $AFG \parallel$ to BC, and from C draw $CG \parallel$ to EF.

Then $FECG$ is a parallelogram.

Now $\triangle AEB = \triangle AEC$,

$\because$ they are on equal bases and between the same $\parallel s$.   I. 38.

$\therefore \triangle ABC$ is double of $\triangle AEC$.

But $\square FECG$ is double of $\triangle AEC$,

$\because$ they are on same base and between same $\parallel s$.   I. 41.

$\therefore \square FECG = \triangle ABC$;   Ax. 6.

and $\square FECG$ has one of its $\angle s$, $CEF = \angle D$.

$\therefore \square FECG$ has been described as was reqd.

Q. E. F.

Ex. 1. Describe a triangle, which shall be equal to a given parallelogram, and have one of its angles equal to a given rectilineal angle.

Ex. 2. Construct a parallelogram, equal to a given triangle, and such that the sum of its sides shall be equal to the sum of the sides of the triangle.

Ex. 3. The perimeter of an isosceles triangle is greater than the perimeter of a rectangle, which is of the same altitude with, and equal to, the given triangle.

PROPOSITION XLIII. THEOREM.

The complements of the parallelograms, which are about the diameter of any parallelogram, are equal to one another.

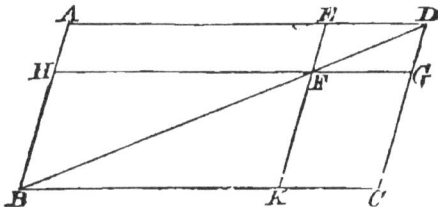

Let $ABCD$ be a $\square$, of which BD is a diagonal, and EG, HK the $\square$s about BD, that is, through which BD passes,

and AF, FC the other $\square$s, which make up the whole figure $ABCD$,

and which are ∴ called the Complements.

Then must complement AF = complement FC.

For ∵ BD is a diagonal of $\square$ AC,

∴ △ ABD = △ CDB ;   I. 34.

and ∵ BF is a diagonal of $\square$ HK,

∴ △ HBF = △ KFB ;   I. 34.

and ∵ FD is a diagonal of $\square$ EG,

∴ △ EFD = △ GDF.   I. 34.

Hence sum of △s HBF, EFD = sum of △s KFB, GDF.

Take these equals from △s ABD, CDB respectively,

then remaining $\square$ AF = remaining $\square$ FC.   Ax. 3.

Q. E. D.

Ex. 1. If through a point O, within a parallelogram $ABCD$, two straight lines are drawn parallel to the sides, and the parallelograms OB, OD are equal ; the point O is in the diagonal AC.

Ex. 2. $ABCD$ is a parallelogram, AMN a straight line meeting the sides BC, CD (one of them being produced) in M, N. Shew that the triangle MBN is equal to the triangle MDC.

PROPOSITION XLIV. PROBLEM.

*To a given straight line to apply a parallelogram, which
shall be equal to a given triangle, and have one of its angles
equal to a given angle.*

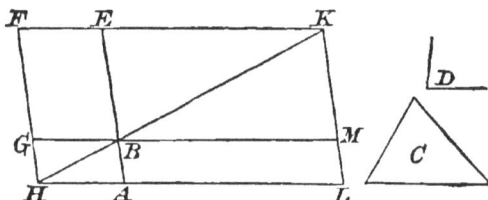

Let AB be the given st. line, C the given △, D the
given ∠.

It is required to apply to AB a □ $= △ C$ *and having one
of its* ∠ s $= ∠ D$.

Make a □ $= △ C$, and having one of its angles $= ∠ D$,   I. 42.
and suppose it to be removed to such a position that one of
the sides containing this angle is in the same st. line with AB,
and let the □ be denoted by $BEFG$.

Produce FG to H, draw $AH \parallel$ to BG or EF, and join BH.

Then ∵ FH meets the ‖s AH, EF,

∴ sum of ∠ s AHF, $HFE = $ two rt. ∠ s ;   I. 29.

∴ sum of ∠ s BHG, HFE is less than two rt. ∠ s ;

∴ HB, FE will meet if produced towards B, E.   Post. 6.

Let them meet in K.

Through K draw $KL \parallel$ to EA or FH,

and produce HA, GB to meet KL in the pts. L, M.

Then $HFKL$ is a □, and HK is its diagonal ;

and AG, ME are □s about HK,

∴ complement $BL = $ complement BF,   I. 43.

∴ □ $BL = △ C$.

Also the □ BL has one of its ∠ s, $ABM = ∠ EBG$, and
. equal to ∠ D.

PROPOSITION XLV.   PROBLEM.

To describe a parallelogram, which shall be equal to a given rectilinear figure, and have one of its angles equal to a given angle.

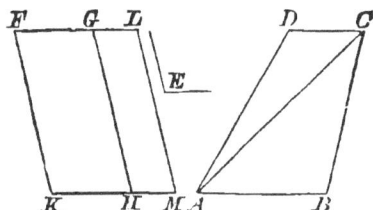

Let $ABCD$ be the given rectil. figure, and E the given $\angle$.

It is required to describe a $\square = $ to $ABCD$, having one of its $\angle$ s $= \angle E$.

Join AC.

Describe a $\square\ FGHK = \triangle ABC$, having $\angle FKH = \angle E$.

I. 42.

To GH apply a $\square\ GHML = \triangle CDA$, having $\angle GHM = \angle E$.

I. 44.

Then $FKML$ is the $\square$ reqd.

For $\because \angle GHM$ and $\angle FKH$ are each $= \angle E$;

$\therefore \angle GHM = \angle FKH$,

$\therefore$ sum of $\angle$ s GHM, $GHK =$ sum of $\angle$ s FKH, GHK

$=$ two rt. $\angle$ s ;   I. 29.

$\therefore KHM$ is a st. line.   I. 14.

Again, $\because HG$ meets the $\|$s FG, KM,

$\angle FGH = \angle GHM$,

$\therefore$ sum of $\angle$ s FGH, $LGH =$ sum of $\angle$ s GHM, LGH

$=$ two rt. $\angle$ s ;   I. 29.

$\therefore FGL$ is a st. line.   I. 14.

Then $\because KF$ is $\|$ to HG, and HG is $\|$ to LM

$\therefore KF$ is $\|$ to LM ;   I. 30.

and KM has been shewn to be $\|$ to FL,

$\therefore FKML$ is a parallelogram,

and $\because FH = \triangle ABC$, and $GM = \triangle CDA$,

$\therefore \square\ FM =$ whole rectil. fig. $ABCD$,

and $\square\ FM$ has one of its $\angle$ s, $FKM = \angle E$.

In the same way a $\square$ may be constructed equal to a given rectil. fig. of any number of sides, and having one of its angles equal to a given angle.   Q. E. F.

1. If one diagonal of a quadrilateral bisect the other, it divides the quadrilateral into two equal triangles.

2. If from any point in the diagonal, or the diagonal produced, of a parallelogram, straight lines be drawn to the opposite angles, they will cut off equal triangles.

3. In a trapezium the straight line, joining the middle points of the parallel sides, bisects the trapezium.

4. The diagonals AC, BD of a parallelogram intersect in O, and P is a point within the triangle AOB; prove that the difference of the triangles CPD, APD is equal to the sum of the triangles APC, BPD.

5. If either diagonal of a parallelogram be equal to a side of the figure, the other diagonal shall be greater than any side of the figure.

6. If through the angles of a parallelogram four straight lines be drawn parallel to its diagonals, another parallelogram will be formed, the area of which will be double that of the original parallelogram.

7. If two triangles have two sides respectively equal and the included angles supplemental, the triangles are equal.

8. Bisect a given triangle by a straight line drawn from a given point in one of the sides.

9. The base AB of a triangle ABC is produced to a point D such that BD is equal to AB, and straight lines are drawn from A and D to E, the middle point of BC; prove that the triangle ADE is equal to the triangle ABC.

10. Prove that a pair of the diagonals of the parallelograms, which are about the diameter of any parallelogram, are parallel to each other.

Proposition XLVI.   Problem.

To describe a square upon a given straight line.

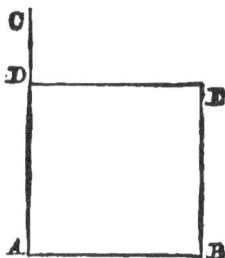

Let AB be the given st. line.

It is required to describe a square on AB.

From A draw $AC \perp$ to AB.   I. 11. Cor.

In AC make $AD = AB$.

Through D draw $DE \parallel$ to AB.   I. 31.

Through B draw $BE \parallel$ to AD.   I. 31.

Then AE is a parallelogram,

and $\therefore AB = ED$, and $AD = BE$.   I. 34.

But $AB = AD$;

$\therefore AB, BE, ED, DA$ are all equal;

$\therefore AE$ is equilateral.

And $\angle BAD$ is a right angle.

$\therefore AE$ is a square,   Def. xxx.

and it is described on AB.

Q. E. F.

Ex. 1. Shew how to construct a rectangle whose sides are equal to two given straight lines.

Ex. 2. Shew that the squares on equal straight lines are equal.

Ex. 3. Shew that equal squares must be on equal straight lines.

Note. The theorems in Ex. 2 and 3 are assumed by Euclid in the proof of Prop. XLVIII.

Proposition XLVII. Theorem.

In any right-angled triangle the square which is described on the side subtending the right angle is equal to the squares described on the sides which contain the right angle.

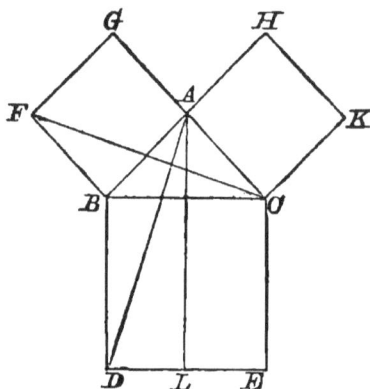

Let ABC be a right-angled $\triangle$, having the rt. $\angle BAC$.
Then must sq. on $BC =$ sum of sqq. on BA, AC.
On BC, CA, AB descr. the sqq. $BDEC$, $CKHA$, $AGFB$.
Through A draw $AL \parallel$ to BD or CE, and join AD, FC.
Then $\because \angle BAC$ and $\angle BAG$ are both rt. $\angle$ s,
$\therefore CAG$ is a st. line ;   I. 14.
and $\because \angle BAC$ and $\angle CAH$ are both rt. $\angle$ s ;
$\therefore BAH$ is a st. line.   I. 14.
Now $\because \angle DBC = \angle FBA$, each being a rt. $\angle$,
adding to each $\angle ABC$, we have
$\angle ABD = \angle FBC$.   Ax. 2.
Then in $\triangle$ s ABD, FBC,
$\because AB = FB$, and $BD = BC$, and $\angle ABD = \angle FBC$,
$\therefore \triangle ABD = \triangle FBC$.   I. 4.
Now $\square BL$ is double of $\triangle ABD$, on same base BD and
between same $\parallel$ s AL, BD.   I. 41.
and sq. BG is double of $\triangle FBC$, on same base FB and between same $\parallel$ s FB, GC ;   I. 41.
$\therefore \square BL =$ sq. BG.

Similarly, by joining AE, BK it may be shewn that
$$\square\ CL = \text{sq. } AK.$$
Now sq. on BC = sum of $\square$ BL and $\square$ CL,
$$= \text{sum of sq. } BG \text{ and sq. } AK,$$
$$= \text{sum of sqq. on } BA \text{ and } AC.$$

Q. E. D.

Ex. 1. Prove that the square, described upon the diagonal of any given square, is equal to twice the given square.

Ex. 2. Find a line, the square on which shall be equal to the sum of the squares on three given straight lines.

Ex. 3. If one angle of a triangle be equal to the sum of the other two, and one of the sides containing this angle being divided into four equal parts, the other contains three of those parts; the remaining side of the triangle contains five such parts.

Ex. 4. The triangles ABC, DEF, having the angles ACB, DFE right angles, have also the sides AB, AC equal to DE, DF, each to each; shew that the triangles are equal in every respect.

NOTE. This Theorem has been already deduced as a Corollary from Prop. E, page 43.

Ex. 5. Divide a given straight line into two parts, so that the square on one part shall be double of the square on the other.

Ex. 6. If from one of the acute angles of a right-angled triangle a line be drawn to the opposite side, the squares on that side and on the line so drawn are together equal to the sum of the squares on the segment adjacent to the right angle and on the hypotenuse.

Ex. 7. In any triangle, if a line be drawn from the vertex at right angles to the base, the difference between the squares on the sides is equal to the difference between the squares on the segments of the base.

PROPOSITION XLVIII. THEOREM.

... described upon one of the sides of a triangle be ... squares described upon the other two sides of it, the ... contained by those sides is a right angle.

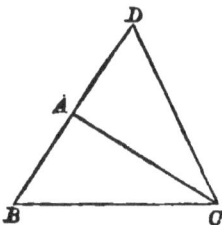

Let the sq. on BC, a side of $\triangle ABC$, be equal to the sum of the sqq. on AB, AC.

Then must $\angle BAC$ be a rt. angle.

From pt. A draw $AD \perp$ to AC.   I. 11.

Make $AD = AB$, and join DC.

Then $\because AD = AB$,

$\therefore$ sq. on $AD =$ sq. on AB;   I. 46, Ex. 2.

add to each sq. on AC.

then sum of sqq. on AD, $AC =$ sum of sqq. on AB, AC.

But $\because \angle DAC$ is a rt. angle,

$\therefore$ sq. on $DC =$ sum of sqq. on AD, AC;   I. 47.

and, by hypothesis,

sq. on $BC =$ sum of sqq. on AB, AC;

$\therefore$ sq. on $DC =$ sq. on BC;

$\therefore DC = BC$.   I. 46, Ex. 3.

Then in $\triangle$s ABC, ADC,

$\because AB = AD$, and AC is common, and $BC = DC$,

$\therefore \angle BAC = \angle DAC$;   I. c.

and $\angle DAC$ is a rt. angle, by construction;

$\therefore \angle BAC$ is a rt. angle.

Q. E. D.

BOOK II.

INTRODUCTORY REMARKS.

THE geometrical figure with which wo are chiefly concerned in this book is the RECTANGLE. A rectangle is said to be *contained by* any two of its adjacent sides.

Thus if $ABCD$ be a rectangle, it is said to be contained by AB, AD, or by any other pair of adjacent sides.

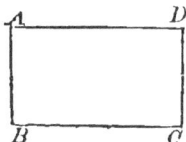

We shall use the abbreviation *rect. AB, AD* to express the words "the rectangle contained by AB, AD."

We shall make frequent use of a Theorem (employed, but not demonstrated, by Euclid) which may be thus stated and proved .

PROPOSITION A.   THEOREM.

If the adjacent sides of one rectangle be equal to the adjacent sides of another rectangle, each to each, the rectangles are equal in area.

Let $ABCD$, $EFGH$ be two rectangles :
and let $AB=EF$ and $BC=FG$.

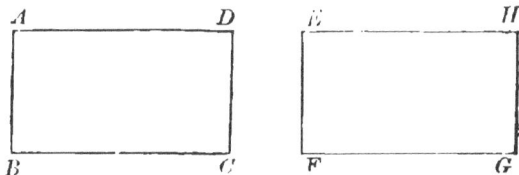

Then must rect. $ABCD=$rect. $EFGH$.

For if the rect. $EFGH$ be applied to the rect. $ABCD$, so that EF coincides with AB,

then FG will fall on BC, $\because \angle EFG = \angle ABC$,

and G will coincide with C, $\because BC=FG$.

Similarly it may be shewn that H will coincide with D,

$\therefore$ rect. $EFGH$ coincides with and is therefore equal to rect $ABCD$.

Q. E. D.

PROPOSITION I. THEOREM.

If there be two straight lines, one of which is divided into any number of parts, the rectangle contained by the two straight lines is equal to the rectangles contained by the undivided line and the several parts of the divided line.

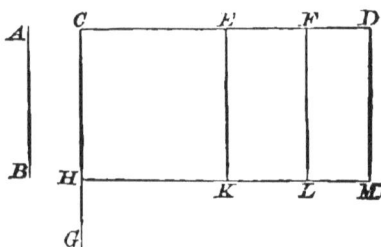

Let AB and CD be two given st. lines,

and let CD be divided into any parts in E, F.

Then must rect. AB, $CD = sum$ of rect. AB, CE and rect. AB, EF and rect. AB, FD.

From C draw $CG \perp$ to CD, and in CG make $CH = AB$.

Through H draw $HM \parallel$ to CD.   I. 31.

Through E, F, and D draw EK, FL, $DM \parallel$ to CH.

Then EK and FL, being each $= CH$, are each $= AB$.

Now $CM =$ sum of CK and EL and FM.

And $CM =$ rect. AB, CD, $\because CH = AB$,

$\quad CK =$ rect. AB, CE, $\because CH = AB$,

$\quad EL =$ rect. AB, EF, $\because EK = AB$,

$\quad FM =$ rect. AB, FD, $\because FL = AB$;

$\therefore$ rect. AB, $CD =$ sum of rect. AB, CE and rect. AB, EF and rect. AB, FD.

Q. E. D.

Ex. If two straight lines be each divided into any number of parts, the rectangle contained by the two lines is equal to the rectangles contained by all the parts of the one taken separately with all the parts of the other.

Proposition II.　Theorem.

If a straight line be divided into any two parts, the rectangles contained by the whole and each of the parts are together equal to the square on the whole line.

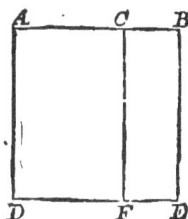

Let the st. line AB be divided into any two parts in C.

Then must

sq. on AB = sum of rect. AB, AC and rect. AB, CB.

On AB describe the sq. $ADEB$.　　　　I. 46.

Through C draw $CF \parallel$ to AD.　　　　I. 31.

Then AE = sum of AF and CE.

Now AE is the sq. on AB,

AF = rect. AB, AC,　　$\because AD = AB$,

CE = rect. AB, CB,　　$\because BE = AB$,

$\therefore$ sq. on AB = sum of rect. AB, AC and rect. AB, CB.

Q. E. D.

Ex. The square on a straight line is equal to four times the square on half the line.

PROPOSITION III. THEOREM.

If a straight line be divided into any two parts, the rectangle contained by the whole and one of the parts is equal to the rect-angle contained by the two parts together with the square on the aforesaid part.

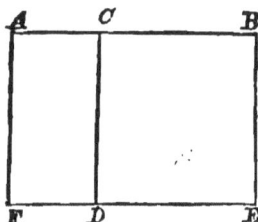

Let the st. line AB be divided into any two parts in C.

Then must

rect. AB, $CB = sum$ *of rect.* AC, CB *and sq. on* CB.

     On CB describe the sq. $CDEB$.   I. 46

From A draw $AF \parallel$ to CD, meeting ED produced in F.

     Then $AE =$ sum of AD and CE.

Now $AE =$ rect. AB, CB, $\because BE = CB$,

     $AD =$ rect. AC, CB, $\because CD = CB$,

     $CE =$ sq. on CB.

$\therefore$ rect. AB, $CB =$ sum of rect. AC, CB and sq. on CB.

Q. E. D.

NOTE. When a straight line is cut in a point, the distances of the point of section from the ends of the line are called the *segments* of the line.

If a line AB be divided in C,

     AC and CB are called the *internal* segments of AB.

If a line AC be produced to B,

     AB and CB are called the *external* segments of AC.

If a straight line be divided into any two parts, the square on the whole line is equal to the squares on the two parts together with twice the rectangle contained by the parts.

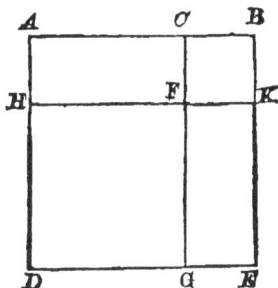

Let the st. line AB be divided into any two parts in C.

Then must

sq. on AB = *sum of sqq.* on AC, CB *and twice rect.* AC, CB.

On AB describe the sq. $ADEB$.   I. 46.

From AD cut off $AH=CB$.   Then $HD=AC$.

Draw $CG \parallel$ to AD, and $HK \parallel$ to AB, meeting CG in F.

Then $\because BK=AH$, $\therefore BK=CB$,   Ax. I.

$\therefore BK$, KF, FC, CB are all equal ; and KBC is a rt. $\angle$;

$\therefore CK$ is the sq. on CB.   Def. xxx.

Also $HG=$ sq. on AC, $\because HF$ and HD each $=AC$.

Now $AE=$ sum of HG, CK, AF, FE,

and $AE=$ sq. on AB,

$HG=$ sq. on AC,

$CK=$ sq. on CB,

$AF=$ rect. AC, CB, $\because CF=CB$,

$FE=$ rect. AC, CB, $\because FG=AC$ and $FK=CB$.

$\therefore$ sq. on $AB=$ sum of sqq. on AC, CB and twice rect. AC, CB.

Q. E. D.

Ex. In a triangle, whose vertical angle is a right angle, a straight line is drawn from the vertex perpendicular to the base. Shew that the rectangle, contained by the segments of the base, is equal to the square on t..e perpendicular.

a E

Proposition V. Theorem.

If a straight line be divided into two equal parts and also into two unequal parts, the rectangle contained by the unequal parts, together with the square on the line between the points of section, is equal to the square on half the line.

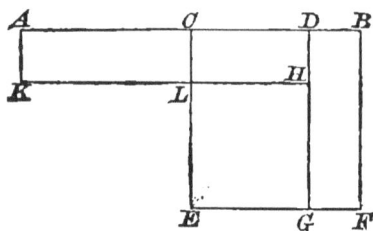

Let the st. line AB be divided equally in C and unequally in D.

Then must

    rect. AD, DB together with sq. on $CD = $ sq. on CB.

On CB describe the sq. $CEFB$.   I. 46.

Draw $DG \parallel$ to CE, and from it cut off $DH = DB$.   I. 31.

Draw $HLK \parallel$ to AD, and $AK \parallel$ to DH.   I. 31.

Then rect. $DF =$ rect. AL, $\because BF = AC$, and $BD = CL$.

     Also $LG =$ sq. on CD, $\because LH = CD$, and $HG = CD$.

Then rect. AD, DB together with sq. on CD

         $= AH$ together with LG

         $=$ sum of AL and CH and LG

         $=$ sum of DF and CH and LG

         $= CF$

         $=$ sq. on CB.

Q. E. D.

PROPOSITION VI. THEOREM.

If a straight line be bisected and produced to any point, the rectangle contained by the whole line thus produced and the part of it produced, together with the square on half the line bisected, is equal to the square on the straight line which is made up of the half and the part produced.

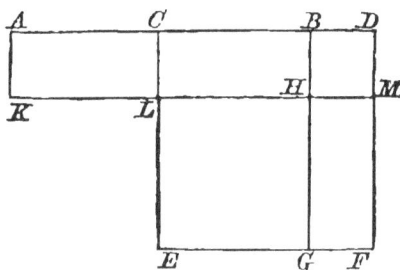

Let the st. line AB be bisected in C and produced to D.

Then must

  rect. AD, DB *together with* sq. *on* $CB=$sq. *on* CD.

    On CD describe the sq. $CEFD$.   I. 46.

   Draw $BG \parallel$ to CE, and cut off $BH = BD$.   I. 31

    Through H draw $KLM \parallel$ to AD I. 31.

    Through A draw $AK \parallel$ to CE.

Now $\because BG = CD$ and $BH = BD$;

      $\therefore HG = CB$;   Ax. 3.

     $\therefore$ rect. $MG =$ rect. AL.   II. A.

Then rect. AD, DB together with sq. on CB

    $=$ sum of AM and LG

    $=$ sum of AL and CM and LG

    $=$ sum of MG and CM and LG

    $= CF$

    $=$ sq. on CD.

              Q. E. D.

Note. We here give the proof of an important theorem, which is usually placed as a corollary to Proposition V.

Proposition B. Theorem.

The difference between the squares on any two straight lines is equal to the rectangle contained by the sum and difference of those lines.

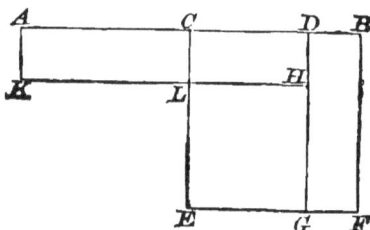

Let AC, CD be two st. lines, of which AC is the greater, and let them be placed so as to form one st. line AD.

Produce AD to B, making $CB = AC$.

Then $AD =$ the sum of the lines AC, CD,

and $DB =$ the difference of the lines AC, CD.

Then must difference between sqq. on AC, $CD = rect.$ AD, DB.

On CB describe the sq. $CEFB$.   I. 46.

Draw $DG \parallel$ to CE, and from it cut off $DH = DB$.   I. 31.

Draw $HLK \parallel$ to AD, and $AK \parallel$ to DH.   I. 31.

Then rect. $DF =$ rect. AL, $\because BF = AC$, and $BD = CL$.

Also $LG =$ sq. on CD, $\because LH = CD$, and $HG = CD$.

Then difference between sqq. on AC, CD

$\qquad$ = difference between sqq. on CB, CD

$\qquad$ = sum of CH and DF

$\qquad$ = sum of CH and AL

$\qquad$ = AH

$\qquad$ = rect. AD, DH

$\qquad$ = rect. AD, DB.

Q. E. D.

Ex. Shew that Propositions V. and VI. might be deduced from this Proposition.

PROPOSITION VII. THEOREM.

If a straight line be divided into any two parts, the squares on the whole line and on one of the parts are equal to twice the rectangle contained by the whole and that part together with the square on the other part.

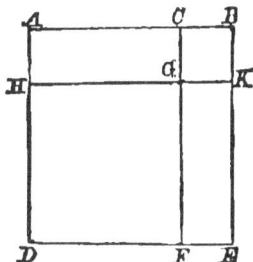

Let AB be divided into any two parts in C.

Then must

sqq. on AB, $BC=$*twice rect.* AB, BC *together with* sq. on AC.

On AB describe the sq. $ADEB$.　　　　　　I. 46.

From AD cut off $AH = CB$.

Draw $CF \parallel$ to AD and $HGK \parallel$ to AB.　　I. 31.

Then $HF=$sq. on AC, and $CK=$sq. on CB.

Then sqq. on AB, $BC=$sum of AE and CK

$=$sum of AK, HF, GE and CK

$=$sum of AK, HF and CE.

Now $AK=$rect. AB, BC,　　$\because BK=BC$;

$CE=$rect. AB, BC,　　$\because BE=AB$;

$HF=$sq. on AC.

$\therefore$ sqq. on AB, $BC=$twice rect. AB, BC together with sq. on AC

Q. E. D.

Ex. If straight lines be drawn from G to B and from G to D, shew that BGD is a straight line.

PROPOSITION VIII. THEOREM.

If a straight line be divided into any two parts, four times the rectangle contained by the whole line and one of the parts, together with the square on the other part, is equal to the square on the straight line which is made up of the whole and the first part.

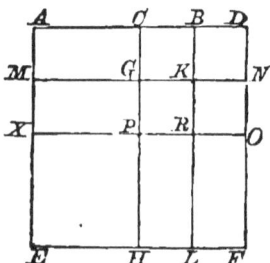

Let the st. line AB be divided into any two parts in C.

Produce AB to D, so that $BD = BC$.

Then must four times rect. AB, BC together with sq. on $C = $ sq. on AD.

On AD describe the sq. $AEFD$.   I. 46.

From AE cut off AM and MX each $= CB$.

Through C, B draw CH, BL ∥ to AE.   I. 31.

Through M, X draw $MGKN$, $XPRO$ ∥ to AD.   I. 31.

Now ∵ $XE = AC$, and $XP = AC$, ∴ $XH =$ sq. on AC.

Also $AG = MP = PL = RF$,   II. A.

and $CK = GR = BN = KO$;   II. A.

∴ sum of these eight rectangles

    = four times the sum of AG, CK

    = four times AK

    = four times rect. AB, BC.

Then four times rect. AB, BC and sq. on AC

    = sum of the eight rectangles and XH

    = $AEFD$

    = sq. on AD.   Q. E. D.

PROPOSITION IX. THEOREM.

If a straight line be divided into two equal, and also into two unequal parts, the squares on the two unequal parts are together double of the square on half the line and of the square on the line between the points of section.

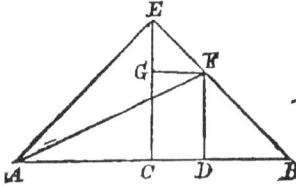

Let AB be divided equally in C and unequally in D.

Then must

sum of sqq. on AD, DB = twice sum of sqq. on AC, CD.

Draw $CE = AC$ at rt. ∠ s to AB, and join EA, EB.

Draw DF at rt. ∠ s to AB, meeting EB in F.

Draw FG at rt. ∠ s to EC, and join AF.

Then ∵ ∠ ACE is a rt. ∠ ,

∴ sum of ∠ s AEC, EAC = a rt. ∠ ;   I. 32.

and ∵ ∠ AEC = ∠ EAC,   I. A.

∴ ∠ AEC = half a rt. ∠ .

So also ∠ BEC and ∠ EBC are each = half a rt. ∠ .

Hence ∠ AEF is a rt. ∠ .

Also, ∵ ∠ GEF is half a rt. ∠ , and ∠ EGF is a rt. ∠ ;

∴ ∠ EFG is half a rt. ∠ ;

∴ ∠ EFG = ∠ GEF, and ∴ $EG = GF$.   I. B. Cor.

So also ∠ BFD is half a rt. ∠ , and $BD = DF$.

Now sum of sqq. on AD, DB

= sq. on AD together with sq. on DF

= sq. on AF I. 47.

= sq. on AE together with sq. on EF I. 47.

= sqq. on AC, EC together with sqq. on EG, GF I. 47.

= twice sq. on AC together with twice sq. on GF

= twice sq. on AC together with twice sq. on CD.

Q. E. D.

PROPOSITION X. THEOREM.

If a straight line be bisected and produced to any point, the square on the whole line thus produced and the square on the part of it produced are together double of the square on half the line bisected and of the square on the line made up of the half and the part produced.

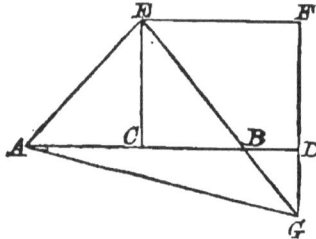

Let the st. line AB be bisected in C and produced to D.
Then must
sum of sqq. on AD, BD = twice sum of sqq. on AC, CD.

Draw $CE \perp$ to AB, and make $CE = AC$.

Join EA, EB and draw $EF \parallel$ to AD and $DF \parallel$ to CE.

Then ∵ ∠s FEB, EFD are together less than two rt. ∠s,

∴ EB and FD will meet if produced towards B, D in some pt. G.

Join AG.

Then ∵ ∠ ACE is a rt. ∠,

∴ ∠s EAC, AEC together = a rt. ∠,

and ∵ ∠ EAC = ∠ AEC,   I. A.

∴ ∠ AEC = half a rt. ∠.

So also ∠s BEC, EBC each = half a rt. ∠.

∴ ∠ AEB is a rt. ∠.

Also ∠ DBG, which = ∠ EBC, is half a rt. ∠,

and ∴ ∠ BGD is half a rt. ∠;

∴ $BD = DG$.   I. B. Cor.

Again, ∵ ∠ FGD = half a rt. ∠, and ∠ EFG is a rt. ∠, I. 34.

∴ ∠ FEG = half a rt. ∠, and $EF = FG$.   I. B. Cor.

Then sum of sqq. on AD, DB

= sum of sqq. on AD, DG

= sq. on AG I. 47.

= sq. on AE together with sq. on EG I. 47.

= sqq. on AC, EC together with sqq. on EF, FG I. 47.

= twice sq. on AC together with twice sq. on EF

= twice sq. on AC together with twice sq. on CD. Q. E. D.

PROPOSITION XI. PROBLEM.

To divide a given straight line into two parts, so that the rectangle contained by the whole and one of the parts shall be equal to the square on the other part.

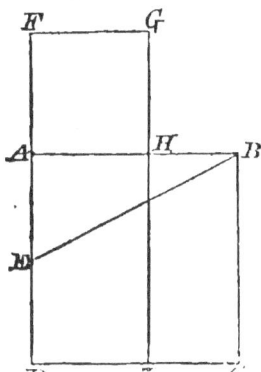

Let AB be the given st. line.

On AB descr. the sq. $ADCB$.	I. 46.
Bisect AD in E and join EB.	I. 10.
Produce DA to F, making $EF = EB$.	
On AF descr. the sq. $AFGH$.	I. 46.

Then AB is divided in H so that rect. AB, $BH = $ sq. on AH.

Produce GH to K.

Then ∵ DA is bisected in E and produced to F,

∴ rect. DF, FA together with sq. on AE	
$=$ sq. on EF	II. 6.
$=$ sq. on EB, ∵ $EB = EF$,	
$=$ sum of sqq. on AB, AE.	I. 47.

Take from each the square on AE.

Then rect. DF, $FA = $ sq. on AB.	Ax. 3.

Now $FK = $ rect. DF, FA, ∵ $FG = FA$.

∴ $FK = AC$.

Take from each the common part AK.

Then $FH = HC$;

that is, sq. on $AH = $ rect. AB, BH, ∵ $BC = AB$.

Thus AB is divided in H as was reqd.

Q. E. F.

Ex. Shew that the squares on the whole line and one of the parts are equal to three times the square on the other part.

PROPOSITION XII.   THEOREM.

In obtuse-angled triangles, if a perpendicular be drawn from either of the acute angles to the opposite side produced, the square on the side subtending the obtuse angle is greater than the squares on the sides containing the obtuse angle, by twice the rectangle contained by the side, upon which, when produced, the perpendicular falls, and the straight line intercepted without the triangle between the perpendicular and the obtuse angle.

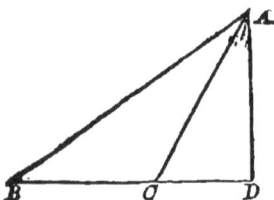

Let ABC be an obtuse-angled $\triangle$, having $\angle ACB$ obtuse.

From A draw $AD \perp$ to BC produced.

Then must sq. on AB be greater than sum of sqq. on BC, CA by twice rect. BC, CD.

For since BD is divided into two parts in C,

sq. on BD=sum of sqq. on BC, CD, and twice rect. BC, CD.

II. 4.

Add to each sq. on DA : then

sum of sqq. on BD, DA=sum of sqq. on BC, CD, DA and twice rect. BC, CD.

Now sqq. on BD, DA=sq. on AB,   I. 47.

and sqq. on CD, DA=sq. on CA ;   I. 47.

$\therefore$ sq. on AB=sum of sqq. on BC, CA and twice rect. BC, CD.

$\therefore$ sq. on AB is greater than sum of sqq. on BC, CA by twice rect. BC, CD.

Q. E. D.

Ex. The squares on the diagonals of a trapezium are together equal to the squares on its two sides, which are not parallel, and twice the rectangle contained by the sides. which are parallel.

PROPOSITION XIII. THEOREM.

In every triangle, the square on the side subtending any of the acute angles is less than the squares on the sides containing that angle, by twice the rectangle contained by either of these sides and the straight line intercepted between the perpendicular, let fall upon it from the opposite angle, and the acute angle.

FIG. 1.   FIG. 2.

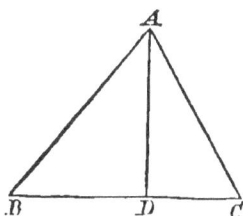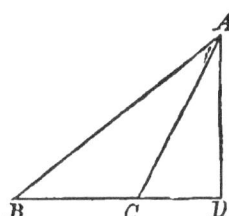

Let *ABC* be any △, having the ∠ *ABC* acute.

From *A* draw *AD* ⊥ to *BC* or *BC* produced.

Then must sq. on AC be less than the sum of sqq. on AB, BC, by twice rect. BC, BD.

For in Fig. 1 *BC* is divided into two parts in *D*,
and in Fig. 2 *BD* is divided into two parts in *C*;

∴ in both cases

sum of sqq. on *BC, BD*=sum of twice rect. *BC, BD* and sq. on *CD*.   II. 7.

Add to each the sq. on *DA*, then

sum of sqq. on *BC, BD, DA*=sum of twice rect. *BC, BD* and sqq. on *CD, DA* ;

∴ sum of sqq. on *BC, AB*=sum of twice rect. *BC, BD* and sq. on *AC*;   I. 47.

∴ sq. on *AC* is less than sum of sqq. on *AB, BC* by twice rect. *BC, BD*.

The case, in which the perpendicular *AD* coincides with *AC*, needs no proof.

Q. E. D.

Ex. Prove that the sum of the squares on any two sides of a triangle is equal to twice the sum of the squares on half the base and on the line joining the vertical angle with the middle point of the base.

PROPOSITION XIV.   PROBLEM.

To describe a square that shall be equal to a given rectilinear figure.

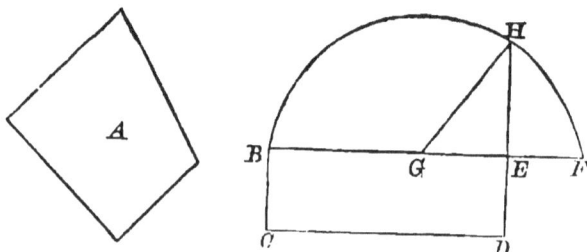

Let A be the given rectil. figure.

It is reqd. to describe a square that shall $= A$.

Describe the rectangular $\square$ $BCDE = A$.   I. 45.

Then if $BE = ED$ the $\square$ $BCDE$ is a square,
          and what was reqd. is done.

But if BE be not $= ED$, produce BE to F, so that $EF = ED$.
          Bisect BF in G; and with centre G and distance GB,
                    describe the semicircle BHF.
          Produce DE to H and join GH.

Then, $\because$ BF is divided equally in G and unequally in E,
          $\therefore$ rect. BE, EF together with sq. on GE
                    $=$ sq. on GF II. 5.
                    $=$ sq. on GH
                    $=$ sum of sqq. on EH, GE.   I. 47.

Take from each the square on GE.

          Then rect. BE, $EF = $ sq. on EH.

But rect. BE, $EF = BD$, $\because$ $EF = ED$;

          $\therefore$ sq. on $EH = BD$;

          $\therefore$ sq. on $EH = $ rectil. figure A.

                              Q. E. F.

1. In a triangle, whose vertical angle is a rignt angle, a straight line is drawn from the vertex perpendicular to the base ; shew that the square on either of the sides adjacent to the right angle is equal to the rectangle contained by the base and the segment of it adjacent to that side.

2. The squares on the diagonals of a parallelogram are together equal to the squares on the four sides.

3. If *ABCD* be any rectangle, and *O* any point either within or without the rectangle, shew that the sum of the squares on *OA*, *OC* is equal to the sum of the squares on *OB*, *OD*.

4. If either diagonal of a parallelogram be equal to one of the sides about the opposite angle of the figure, the square on it shall be less than the square on the other diameter, by twice the square on the other side about that opposite angle.

5. Produce a given straight line *AB* to *C*, so that the rectangle, contained by the sum and difference of *AB* and *AC*, may be equal to a given square.

6. Shew that the sum of the squares on the diagonals of any quadrilateral is less than the sum of the squares on the four sides, by four times the square on the line joining the middle points of the diagonals.

7. If the square on the perpendicular from the vertex of a triangle is equal to the rectangle, contained by the segments of the base, the vertical angle is a right angle.

8. If two straight lines be given, shew how to produce one of them so that the rectangle contained by it and the produced part may be equal to the square on the other.

9. If a straight line be divided into three parts, the square on the whole line is equal to the sum of the squares on the parts together with twice the rectangle contained by each two of the parts.

10. In any quadrilateral the squares on the diagonals are together equal to twice the sum of the squares on the straight lines joining the middle points of opposite sides.

11. If straight lines be drawn from each angle of a triangle to bisect the opposite sides, four times the sum of the squares on these lines is equal to three times the sum of the squares on the sides of the triangle.

12. *CD* is drawn perpendicular to *AB*, a side of the triangle *ABC*, in which *AC*=*AB*. Shew that the square on *CD* is equal to the square on *BD* together with twice the rectangle *AD*, *DB*.

13. The hypotenuse *AB* of a right-angled triangle *ABC* is trisected in the points *D*, *E*; prove that if *CD*, *CE* be joined, the sum of the squares on the sides of the triangle *CDE* is equal to two-thirds of the square on *AB*.

14. The square on the hypotenuse of an isosceles right angled triangle is equal to four times the square on the perpendicular from the right angle on the hypotenuse.

15. Divide a given straight line into two parts, so that the rectangle contained by them shall be equal to the square described upon a straight line, which is less than half the line divided.

NOTE 6.—*On the Measurement of Areas.*

To measure a Magnitude, we fix upon some magnitude of the same kind to serve as a standard or unit ; and then any magnitude of that kind is measured by the number of times it contains this unit, and this number is called the MEASURE of the quantity.

Suppose, for instance, we wish to measure a straight line AB. We take another straight line EF for our standard,

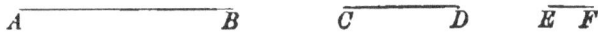

$$\overline{A \qquad\qquad B} \quad \overline{C \qquad D} \quad \overline{E \; F}$$

and then we say

if AB contain EF three times, the measure of AB is **3**,
if .... four..**4**,
if .... x ....x.

Next suppose we wish to measure two straight lines AB, CD by the same standard EF.

If AB contain EF m times
and CD .... n times,

where m and n stand for numbers, whole or fractional, we say that AB and CD are *commensurable.*

But it may happen that we may be able to find a standard line EF, such that it is contained an exact number of times in AB ; and yet there is no number, whole or fractional, which will express the number of times EF is contained in CD.

In such a case, where no unit-line can be found, such that it is contained an exact number of times in *each* of two lines AB, CD, these two lines are called *incommensurable.*

In the processes of Geometry we constantly meet with incommensurable magnitudes. Thus the side and diagonal of a square are incommensurables ; and so are the diameter and circumference of a circle.

Next, suppose two lines AB, AC to be at right angles to each other and to be commensurable, so that AB contains four times a certain unit of linear measurement, which is contained by AC three times.

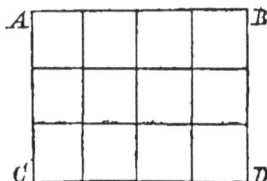

Divide AB, AC into four and three equal parts respectively, and draw lines through the points of division parallel to AC, AB respectively ; then the rectangle $ACDB$ is divided into a number of equal squares, each constructed on a line equal to the unit of linear measurement.

If one of these squares be taken as the unit of area, the *measure* of the area of the rectangle $ACDB$ will be the number of these squares.

Now this number will evidently be the same as that obtained by multiplying the measure of AB by the measure of AC; that is, the measure of AB being 4 and the measure of AC 3, the measure of $ACDB$ is 4×3 or 12. (Algebra, Art. 38.)

And *generally*, if the measures of two adjacent sides of a rectangle, supposed to be commensurable, be a and b, then the measure of the rectangle will be ab. (Algebra, Art. 39.)

If all lines were commensurable, then, whatever might be the length of two adjacent sides of a rectangle, we might select the unit of length, so that the measures of the two sides should be whole numbers ; and then we might apply the processes of Algebra to establish many Propositions in Geometry by simpler methods than those adopted by Euclid.

Take, for example, the theorem in Book II. Prop. IV.

If all lines were commensurable we might proceed thus :—

Let the measure of AC be x,

.... of CB ... y,

Then the measure of AB is $x+y$.

Now $(x + y)^2 = x^2 + y^2 + 2xy$,

which proves the theorem.

But, inasmuch as all lines are not commensurable, we have in Geometry to treat of *magnitudes* and not of *measures:* that is, when we use the symbol A to represent a line (as in I. 22), A stands for the line itself and not, as in Algebra, for the number of units of length contained by the line.

The method, adopted by Euclid in Book II. to explain the relations between the rectangles contained by certain lines, is more exact than any method founded upon Algebraical principles can be ; because his method applies not merely to the case in which the sides of a rectangle are commensurable, but also to the case in which they are incommensurable.

The student is now in a position to understand the practical application of the theory of Equivalence of Areas, of which the foundation is the 35th Proposition of Book I. We shall give a few examples of the use made of this theory in Mensuration.

Area of a Parallelogram.

The area of a parallelogram $ABCD$ is equal to the area of the rectangle $ABEF$ on the same base AB and between the same parallels AB, FC.

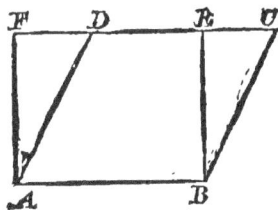

Now BE is the altitude of the parallelogram $ABCD$ if AB be taken as the base.

Hence area of $\square$ $ABCD$ = rect. AB, BE.

If then the measure of the base be denoted by b,

and .... altitude .... h,

the measure of the area of the $\square$ will be denoted by bh.

That is, when the base and altitude are commensurable,

measure of area = measure of base into measure of altitude.

R. P.

Area of a Triangle.

If from one of the angular points A of a triangle ABC, a perpendicular AD be drawn to BC, Fig. 1, or to BC produced, Fig. 2,

Fɪɢ. 1.   Fɪɢ. 2.

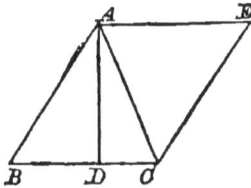 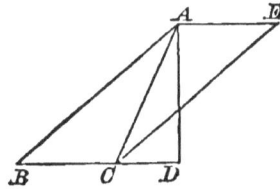

and if, in both cases, a parallelogram $ABCE$ be completed of which AB, BC are adjacent sides,

area of $\triangle ABC =$ half of area of $\square ABCE$.

Now if the measure of BC be b,

and ....AD... h,

measure of area of $\square ABCE$ is bh ;

∴ measure of area of $\triangle ABC$ is $\dfrac{bh}{2}$.

Area of a Rhombus.

Let $ABCD$ be the given rhombus.

Draw the diagonals AC and BD, cutting one another in O.

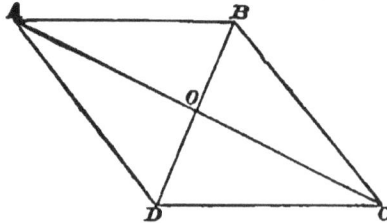

It is easy to prove that AC and BD bisect each other at right angles.

Then if the measure of AC be x,

and .... BD ... y,

measure of area of rhombus $=$ twice measure of $\triangle ACD$.

$$= \text{twice } \frac{xy}{4}$$

$$= \frac{xy}{2}.$$

Area of a Trapezium.

Let *ABCD* be the given trapezium, having the sides *AB*, *CD* parallel.

Draw *AE* at right angles to *CD*.

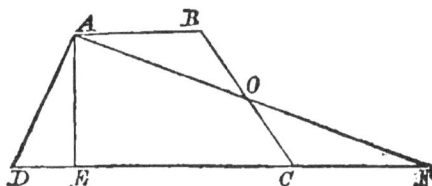

Produce *DC* to *F*, making $CF = AB$.

Join *AF*, cutting *BC* in *O*.

Then in △s *AOB*, *COF*,

$\because \angle BAO = \angle CFO$, and $\angle AOB = \angle FOC$, and $AB = CF$;

$\therefore \triangle COF = \triangle AOB$.　　　　　I. 26.

Hence trapezium $ABCD = \triangle ADF$.

Now suppose the measures of *AB*, *CD*, *AE* to be m, n, p respectively ;

$\therefore$ measure of $DF = m + n$, $\because CF = AB$.

Then measure of area of trapezium

$= \tfrac{1}{2}$ (measure of $DF \times$ measure of AE)

$= \tfrac{1}{2} (m + n) \times p$.

That is, the measure of the area of a trapezium is found by multiplying half the measure of the sum of the parallel sides by the measure of the perpendicular distance between the parallel sides.

Area of an Irregular Polygon.

There are three methods of finding the area of an irregular polygon, which we shall here briefly notice.

I. *The polygon may be divided into triangles,* and the area of each of these triangles be found separately.

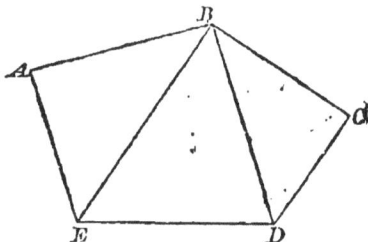

Thus the area of the irregular polygon $ABCDE$ is equal to the sum of the areas of the triangles ABE, EBD, DBC.

II. *The polygon may be converted into a single triangle of equal area.*

If $ABCDE$ be a pentagon, we can convert it into an equivalent quadrilateral by the following process :

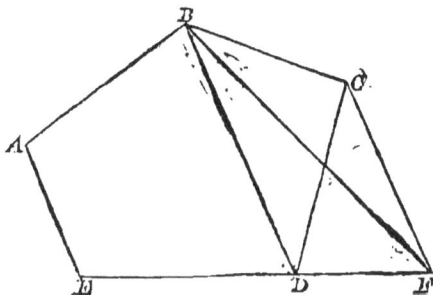

Join BD and draw CF parallel to BD, meeting ED produced in F, and join BF.

Then will quadrilateral $ABFE$ = pentagon $ABCDE$.

For $\triangle BDF = \triangle BCD$, on same base BD and between same parallels.

If, then, from the pentagon we remove $\triangle BCD$, and add $\triangle BDF$ to the remainder, we obtain a quadrilateral $ABFE$ equivalent to the pentagon $ABCDE$.

The quadrilateral may then, by a similar process, be converted into an equivalent triangle, and thus a polygon of any number of sides may be gradually converted into an equivalent triangle.

The area of this triangle may then be found.

III. The third method is chiefly employed in practice by Surveyors

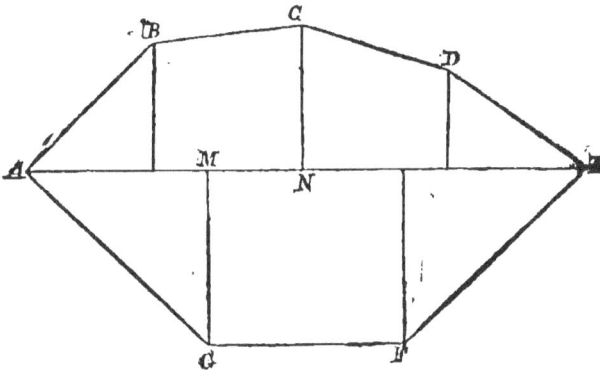

Let *ABCDEFG* be an irregular polygon.

Draw *AE*, the longest diagonal, and drop perpendiculars on *AE* from the other angular points of the polygon.

The polygon is thus divided into figures which are either right-angled triangles, rectangles, or trapeziums ; and the areas of each of these figures may be readily calculated.

Note 7.   *On Projections.*

The projection of a *point B,* on a straight line of unlimited length *AE,* is the point *M* at the foot of the perpendicular dropped from *B* on *AE.*

The projection of a *straight line BC,* on a straight line of unlimited length *AE,* is *MN,*—the part of *AE* intercepted between perpendiculars drawn from *B* and *C.*

When two lines, as *AB* and *AE,* form an angle, the pro· jection of *AB* on *AE* is *AM.*

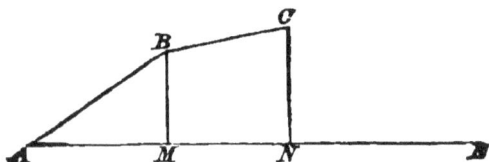

We might employ the term projection with advantage to shorten and make clearer the enunciations of Props. xii. and xiii. of Book II.

Thus the enunciation of Prop. xii. might be :—

" In oblique-angled triangles, the square on the side sub· tending the obtuse angle is greater than the squares on the sides containing that angle, by twice the rectangle contained by one of these sides and the projection of the other on it."

The enunciation of Prop. xiii. might be altered in a similar manner.

NOTE 8.   On Loci.

Suppose we have to determine the position of a point, which is equidistant from the extremities of a given straight line *BC*.

There is an infinite number of points satisfying this con-dition, for the vertex of any isosceles triangle, described on *BC* as its base, is equidistant from *B* and *C*.

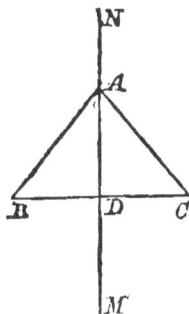

Let *ABC* be *one* of the isosceles triangles described on *BC*.

If *BC* be bisected in *D*, *MN*, a perpendicular to *BC* drawn through *D*, will pass through *A*.

It is easy to shew that any point in *MN*, or *MN* produced in either direction, is equidistant from *B* and *C*.

It may also be proved that no point out of *MN* is equi-distant from *B* and *C*.

The line *MN* is called the Locus of all the points, infinite in number, which are equidistant from *B* and *C*.

DEF. In plane Geometry *Locus* is the name given to a line, straight or curved, all of whose points satisfy a certain geometrical condition (or have a common property), to the exclusion of all other points.

Next, suppose we have to determine the position of a point, which is equidistant from three given points A, B, C, not in the same straight line.

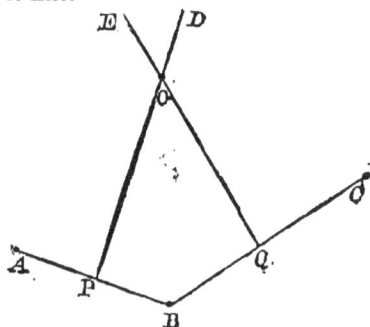

If we join A and B, we know that all points equidistant from A and B lie in the line PD, which bisects AB at right angles.

If we join B and C, we know that all points equidistant from B and C lie in the line QE, which bisects BC at right angles.

Hence O, the point of intersection of PD and QE, is the only point equidistant from A, B and C.

PD is the Locus of points equidistant from A and B,

QE.. B and C,

and the Intersection of these Loci determines the point, which is equidistant from A, B and C.

Examples of Loci.

Find the loci of

(1) Points at a given distance from a given point.

(2) Points at a given distance from a given straight line.

(3) The middle points of straight lines drawn from a given point to a given straight line.

(4) Points equidistant from the arms of an angle.

(5) Points equidistant from a given circle.

(6) Points equally distant from two straight lines which intersect.

NOTE 9. *On the Methods employed in the solution of Problems.*

In the solution of Geometrical Exercises, certain methods may be applied with success to particular classes of questions.

We propose to make a few remarks on these methods, so far as they are applicable to the first two books of Euclid's Elements.

The Method of Synthesis.

In the Exercises, attached to the Propositions in the preceding pages, the construction of the diagram, necessary for the solution of each question, has usually been fully described, or sufficiently suggested.

The student has in most cases been required simply to apply the geometrical fact, proved in the Proposition preceding the exercise, in order to arrive at the conclusion demanded in the question.

This way of proceeding is called Synthesis ($\sigma\acute{\nu}\nu\theta\epsilon\sigma\iota\varsigma$ = composition), because in it we proceed by a regular chain of reasoning from what is *given* to what is *sought*. This being the method employed by Euclid throughout the Elements, we have no need to exemplify it here.

The Method of Analysis.

The solution of many Problems is rendered more easy *by supposing the problem solved and the diagram constructed.* It is then often possible to observe relations between lines, angles and figures in the diagram, which are suggestive of the steps by which the necessary construction might have been effected.

This is called the Method of Analysis ($\acute{\alpha}\nu\acute{\alpha}\lambda\upsilon\sigma\iota\varsigma$ = resolution). It is a method of discovering truth by reasoning concerning things unknown or propositions merely supposed, as if the one were given or the other were really true. The process can best be explained by the following examples.

Our first example of the Analytical process shall be the 31st Proposition of Euclid's First Book.

Ex. 1. *To draw a straight line through a given point parallel to a given straight line.*

Let A be the given point, and BC be the given straight line.

Suppose the problem to be effected, and EF to be the straight line required.

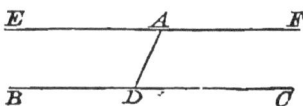

Now we know that any straight line AD drawn from A to meet BC makes equal angles with EF and BC. (I. 29.)

This is a fact from which we can work backward, and arrive at the steps necessary for the solution of the problem ; thus :

Take any point D in BC, join AD, make $\angle EAD = \angle ADC$, and produce EA to F: then EF must be parallel to BC.

Ex. 2. *To inscribe in a triangle a rhombus, having one of its angles coincident with an angle of the triangle.*

Let ABC be the given triangle.

Suppose the problem to be effected, and $DBFE$ to be the rhombus.

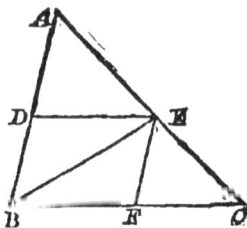

Then if EB be joined, $\angle DBE = \angle FBE$.

This is a fact from which we can work backward, and deduce the necessary construction ; thus :

Bisect $\angle ABC$ by the straight line BE, meeting AC in E.
Draw ED and EF parallel to BC and AB respectively.
Then $DBFE$ is the rhombus required. (See Ex. 4, p. 59.)

Ex. 3. *To determine the point in a given straight line, at which straight lines, drawn from two given points, on the same side of the given line, make equal angles with it.*

Let CD be the given line, and A and B the given points.

Suppose the problem to be effected, and P to be the point required.

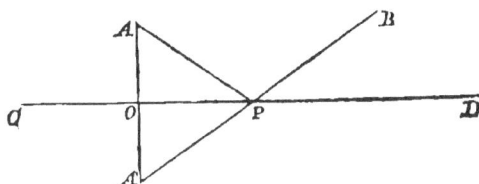

We then reason thus :

If BP were produced to some point A',

$\angle CPA'$, being $= \angle BPD$, will be $= \angle APC$.

Again, if PA' be made equal to PA,

AA' will be bisected by CP at right angles.

This is a fact from which we can work backward, and find the steps necessary for the solution of the problem ; thus :

From A draw $AO \perp$ to CD.

Produce AO to A', making $OA' = OA$.

Join BA', cutting CD in P.

Then P is the point required.

Note 10.   *On Symmetry.*

The problem, which we have just been considering, suggests the following remarks :

If two points, A and A', be so situated with respect to a straight line CD, that CD bisects at right angles the straight line joining A and A', then A and A' are said to be *symmetrical* with regard to CD.

The importance of symmetrical relations, as suggestive of methods for the solution of problems, cannot be fully shewn to a learner, who is unacquainted with the properties of the circle. The following example, however, will illustrate this part of the subject sufficiently for our purpose at present.

Find a point in a given straight line, such that the sum of its distances from two fixed points on the same side of the line is a minimum, that is, less than the sum of the distances of any other point in the line from the fixed points.

Taking the diagram of the last example, suppose *CD* to be the given line, and *A*, *B* the given points.

Now if *A* and *A'* be symmetrical with respect to *CD*, we know that *every* point in *CD* is equally distant from *A* and *A'*. (See Note 8, p. 103.)

Hence the sum of the distances of any point in *CD* from *A* and *B* is equal to the sum of the distances of that point from *A'* and *B*.

But the sum of the distances of a point in *CD* from *A'* and *B* is the least possible when it lies in the straight line joining *A'* and *B*.

Hence the point *P*, *determined as in the last example*, is the point required.

NOTE. Propositions IX., X., XI., XII. of Book I. give good examples of symmetrical constructions.

NOTE 11.   *Euclid's Proof of I. 5.*

The angles at the base of an isosceles triangle are equal to one another ; and if the equal sides be produced, the angles upon the other side of the base shall be equal.

Let *ABC* be an isosceles △, having *AB* = *AC*

Produce *AB*, *AC* to *D* and *E*.

Then must ∠ *ABC* = ∠ *ACB*,

and ∠ *DBC* = ∠ *ECB*.

In *BD* take any pt. *F*.

From *AE* cut off *AG = AF*.

Join *FC* and *GB*.

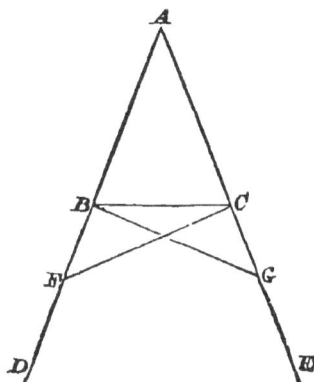

Then in △s *AFC, AGB,*

∵ *FA = GA*, and *AC = AB*, and ∠ *FAC* = ∠ *GAB*,

∴ *FC = GB*, and ∠ *AFC* = ∠ *AGB*, and ∠ *ACF* = ∠ *ABG*.

I. 4.

Again,   ∵ *AF = AG*,

of which the parts *AB, AC* are equal,

∴ remainder *BF* = remainder *CG*.   Ax. 3.

Then in △s *BFC, CGB,*

∵ *BF = CG*, and *FC = GB*, and ∠ *BFC* = ∠ *CGB*,

∴ ∠ *FBC* = ∠ *GCB*, and ∠ *BCF* = ∠ *CBG*,   I. 4.

Now it has been proved that ∠ *ACF* = ∠ *ABG*,

of which the parts ∠ *BCF* and ∠ *CBG* are equal;

∴ remaining ∠ *ACB* = remaining ∠ *ABC*.   Ax. 3.

Also it has been proved that ∠ *FBC* = ∠ *GCB*,

that is,   ∠ *DBC* = ∠ *ECB*.

Q. E. D.

Note 12.   *Euclid's Proof of I. 6.*

If two angles of a triangle be equal to one another, the sides also, which subtend the equal angles, shall be equal to one another.

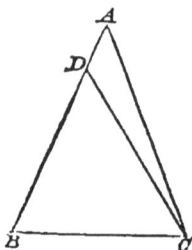

In △ *ABC* let ∠ *ACB* = ∠ *ABC*.

Then must AB = AC.

For if not, *AB* is either greater or less than *AC*.

Suppose *AB* to be greater than *AC*.

From *AB* cut off *BD = AC*, and join *DC*.

Then in △s *DBC*, *ACB*,

∵ *DB = AC*, and *BC* is common, and ∠ *DBC* = ∠ *ACB*,

∴ △ *DBC* = △ *ACB* ;   I. 4.

that is, the less = the greater ; which is absurd.

∴ *AB* is not greater than *AC*.

Similarly it may be shewn that *AB* is not less than *AC*;

∴ *AB = AC*.

Q. E. D.

Note 13.   *Euclid's Proof of I. 7.*

Upon the same base and on the same side of it, there cannot be two triangles that have their sides which are terminated in one extremity of the base equal to one another, and their sides which are terminated in the other extremity of the base equal also.

If it be possible, on the same base *AB*, and on the same side of it, let there be two △s *ACB*, *ADB*, such that *AC = AD*, and also *BC = BD*.

Join *CD*.

First, when the vertex of each of the △s is *outside* the other △ (Fig. 1.) ;

Fɪɢ 1.   Fɪɢ. 2.

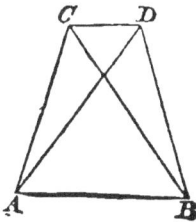 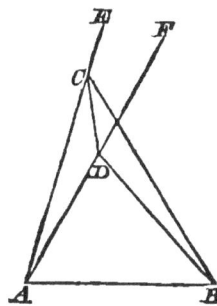

$$\because AD = AC,$$
$$\therefore \angle ACD = \angle ADC. \qquad \text{I. 5.}$$

But $\angle ACD$ is greater than $\angle BCD$;

$$.. \angle ADC \text{ is greater than } \angle BCD ;$$

much more is $\angle BDC$ greater than $\angle BCD$.

Again, $\because BC = BD,$

$$\therefore \angle BDC = \angle BCD,$$

that is, $\angle BDC$ is both equal to and greater than $\angle BCD$; which is absurd.

Secondly, when the vertex D of one of the △s falls *within* the other △ (Fig. 2) ;

Produce AC and AD to E and F

Then $\because AC = AD.$

$$\therefore \angle ECD = \angle FDC. \qquad \text{I. 5.}$$

But $\angle ECD$ is greater than $\angle BCD$;

$$\therefore \angle FDC \text{ is greater than } \angle BCD ;$$

much more is $\angle BDC$ greater than $\angle BCD$.

Again, $\because BC = BD,$

$$\therefore \angle BDC = \angle BCD ;$$

that is, $\angle BDC$ is both equal to and greater than $\angle BCD$: which is absurd.

Lastly, when the vertex D of one of the △s falls on a side BC of the other, it is plain that BC and BD cannot be equal.   Q. E. D.

Note 14.   *Euclid's Proof of I.* 8.

If two triangles have two sides of the one equal to two sides of the other, each to each, and have likewise their bases equal, the angle which is contained by the two sides of the one must be equal to the angle contained by the two sides of the other.

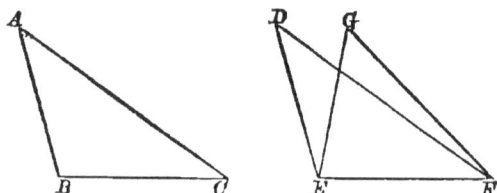

Let the sides of the △ s *ABC, DEF* be equal, each to each, that is, *AB=DE, AC=DF* and *BC=EF*.

Then must ∠ *BAC=* ∠ *EDF*.

Apply the △ *ABC* to the △ *DEF*.
        so that pt. *B* is on pt. *E*, and *BC* on *EF*.
Then   ∵ *BC=EF*,
        ∴ *C* will coincide with *F*,
        and *BC* will coincide with *EF*.

Then *AB* and *AC* must coincide with *DE* and *DF*.

For if *AB* and *AC* have a different position, as *GE, GF*, then upon the same base and upon the same side of it there can be two △ s, which have their sides which are terminated in one extremity of the base equal, and their sides which are terminated in the other extremity of the base also equal : which is impossible.   I. 7.

∴ since base *BC* coincides with base *EF*,

   AB must coincide with *DE*, and *AC* with *DF* ;

   ∴ ∠ *BAC* coincides with and is equal to ∠ *EDF*.

                                          Q. E. D.

NOTE 15. *Another Proof of I. 24.*

In the △s ABC, DEF, let $AB = DE$ and $AC = DF$, and let ∠ BAC be greater than ∠ EDF.

Then must BC be greater than EF.

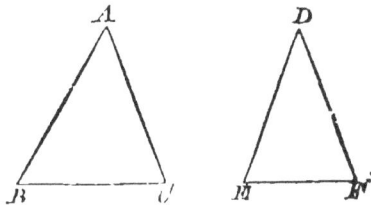

Apply the △ DEF to the △ ABC
so that DE coincides with AB.
Then ∵ ∠ EDF is less than ∠ BAC,
DF will fall between BA and AC,
and F will fall *on*, or *above*, or *below*, BC.

I. If F fall on BC,

BF is less than BC;

∴ EF is less than BC.

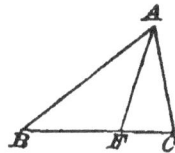

II. If F fall *above* BC,

BF, FA together are less than BC, CA,

and $FA = CA$;

∴ BF is less than BC;

∴ EF is less than BC.

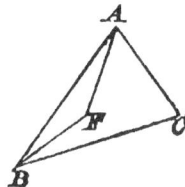

III. If F fall *below* BC,

let AF cut BC in O.

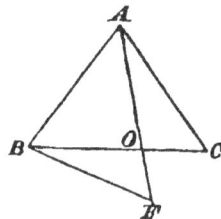

Then BO, OF together are greater than BF,   I. 20.
and OC, AO .... AC ;   I. 20.
∴ BC, AF .... BF, AC together,
        and $AF = AC$,
    ∴ BC is greater than BF ·
and ∴ EF is less than BC.   Q. E. D.

8

Note 16. *Euclid's Proof of I. 26.*

If two triangles have two angles of the one equal to two angles of the other, each to each, and one side equal to one side, viz., either the sides adjacent to the equal angles, or the sides opposite to equal angles in each; then shall the other sides be equal, each to each; and also the third angle of the one to the third angle of the other.

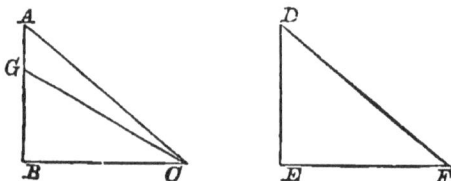

In △s ABC, DEF,

Let ∠ ABC = ∠ DEF, and ∠ ACB = ∠ DFE;
and *first,*

Let tho sides adjacent to the equal ∠ s in each bo equal,
that is, lct $BC=EF$.

Then must $AB=DE$, and $AC=DF$, and ∠ BAC = ∠ EDF.

For if AB be not$=DE$, ono of them must bc the greater.

Let AB bo the greater, and make $GB=DE$, and join GC

Then in △s GBC, DEF,

∵ $GB=DE$, and $BC=EF$, and ∠ GBC = ∠ DEF,

∴ ∠ GCB = ∠ DFE.   I. 4.

But   ∠ ACB = ∠ DFE by hypothesis;

∴ ∠ GCB = ∠ ACB;

that is, the less$=$the greater, which is impossible.

∴ AB is not greater than DE.

In the same way it may bo shewn that AB is not less than DE;

∴ $AB=DE$.

Then in △s ABC, DEF,

∵ $AB=DE$, and $BC=EF$, and ∠ ABC = ∠ DEF,

∴ $AC=DF$. and ∠ BAC = ∠ EDF.   I. 4.

Next, let the sides which are opposite to equal angles in each triangle be equal, viz., $AB=DE$.

Then must $AC=DF$, *and* $BC=EF$, *and* $\angle BAC = \angle EDF$.

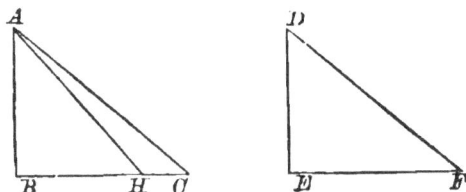

For if BC be not $=EF$, let BC be the greater, and make $BH=EF$, and join AH.

Then in $\triangle$s ABH, DEF,

∵ $AB=DE$, and $BH=EF$, and $\angle ABH = \angle DEF$,

∴ $\angle AHB = \angle DFE$.   I. 4.

But $\angle ACB = \angle DFE$, by hypothesis,

∴ $\angle AHB = \angle ACB$;

that is, the exterior $\angle$ of $\triangle$ AHC is equal to the interior and opposite $\angle$ ACB, which is impossible.

∴ BC is not greater than EF.

In the same way it may be shewn that BC is not less than EF ;

∴ $BC=EF$.

Then in $\triangle$s ABC, DEF,

∵ $AB=DE$, and $BC=EF$, and $\angle ABC = \angle DEF$,

∴ $AC=DF$, and $\angle BAC = \angle EDF$.   I. 4.

Q. E. D.

Miscellaneous Exercises on Books I. and II.

1. *AB* and *CD* are equal straight lines, bisecting one another at right angles. Shew that *ACBD* is a square.

2. From a point in the side of a parallelogram draw a line dividing the parallelogram into two equal parts.

3. In the triangle *FDC*, if *FCD* be a right angle, and angle *FDC* be double of angle *CFD*, shew that *FD* is double of *DC*.

4. If *ABC* be an equilateral triangle, and *AD, BE* be perpendiculars to the opposite sides intersecting in *F*; shew that the square on *AB* is equal to three times the square on *AF*.

5. Describe a rhombus, which shall be equal to a given triangle, and have each of its sides equal to one side of the triangle.

6. From a given point, outside a given straight line, draw a line making with the given line an angle equal to a given rectilineal angle.

7. If two straight lines be drawn from two given points to meet in a given straight line, shew that the sum of these lines is the least possible, when they make equal angles with the given line.

8. *ABCD* is a parallelogram, whose diagonals *AC, BD* intersect in *O*; shew that if the parallelograms *AOBP, DOCQ* be completed, the straight line joining *P* and *Q* passes through *O*.

9. *ABCD, EBCF* are two parallelograms on the same base *BC*, and so situated that *CF* passes through *A*. Join *DF*, and produce it to meet *BE* produced in *K*; join *FB*, and prove that the triangle *FAB* equals the triangle *FEK*.

10. The alternate sides of a polygon are produced to meet; shew that all the angles at their points of intersection together with four right angles are equal to all the interior angles of the polygon.

11. Shew that the perimeter of a rectangle is always greater than that of the square equal to the rectangle.

12. Shew that the opposite sides of an equiangular hexagon are parallel, though they be not equal.

13. If two equal straight lines intersect each other anywhere at right angles, shew that the area of the quadrilateral formed by joining their extremities is invariable, and equal to one-half the square on either line.

14. Two triangles ACB, ADB are constructed on the same side of the same base AB. Show that if $AC=BD$ and $AD=BC$, then CD is parallel to AB; but if $AC=BC$ and $AD=BD$, then CD is perpendicular to AB.

15. AB is the hypotenuse of a right-angled triangle ABC: find a point D in AB, such that DB may be equal to the perpendicular from D on AC.

16. Find the locus of the vertices of triangles of equal area on the same base, and on the same side of it.

17. Shew that the perimeter of an isosceles triangle is less than that of any triangle of equal area on the same base.

18. If each of the equal angles of an isosceles triangle be equal to one-fourth the vertical angle, and from one of them a perpendicular be drawn to the base, meeting the opposite side produced, then will the part produced, the perpendicular, and the remaining side, form an equilateral triangle.

19. If a straight line terminated by the sides of a triangle be bisected, shew that no other line terminated by the same two sides can be bisected in the same point.

20. Show how to bisect a given quadrilateral by a straight line drawn from one of its angles.

21. Given the lengths of the two diagonals of a rhombus, construct it.

22. $ABCD$ is a quadrilateral figure : construct a triangle whose base shall be in the line AB, such that its altitude shall be equal to a given line, and its area equal to that of the quadrilateral.

23. If from any point in the base of an isosceles triangle perpendiculars be drawn to the sides, their sum will be equal to the perpendicular from either extremity of the base upon the opposite side.

24. If ABC be a triangle, in which C is a right angle, and DE be drawn from a point D in AC at right angles to AB, prove that the rectangles AB, AE and AC, AD are equal.

25. A line is drawn bisecting parallelogram $ABCD$, and meeting AD, BC in E and F: shew that the triangles EBF, CED are equal.

26. Upon the hypotenuse BC and the sides CA, AB of a right-angled triangle ABC, squares $BDEC$, AF and AG are described: shew that the squares on DG and EF are together equal to five times the square on BC.

27. If from the vertical angle of a triangle three straight lines be drawn, one bisecting the angle, the second bisecting the base, and the third perpendicular to the base, shew that the first lies, both in position and magnitude, between the other two.

28. If ABC be a triangle, whose angle A is a right angle, and BE, CF be drawn bisecting the opposite sides respectively, shew that four times the sum of the squares on BE and CF is equal to five times the square on BC.

29. Let ACB, ADB be two right-angled triangles having a common hypotenuse AB. Join CD and on CD produced both ways draw perpendiculars AE, BF. Shew that the sum of the squares on CE and CF is equal to the sum of the squares on DE and DF.

30. In the base AC of a triangle take any point D: bisect AD, DC, AB, BC at the points E, F, G, H respectively. Shew that EG is equal and parallel to FH.

31. If AD be drawn from the vertex of an isosceles triangle ABC to a point D in the base, shew that the rectangle BD, DC is equal to the difference between the squares on AB and AD.

32. If in the sides of a square four points be taken at equal distances from the four angular points taken in order, the figure contained by the straight lines, which join them, shall also be a square.

33. If the sides of an equilateral and equiangular pentagon be produced to meet, shew that the sum of the angles at the points of meeting is equal to two right angles.

34. Describe a square that shall be equal to the difference between two given and unequal squares.

35. $ABCD$, $AECF$ are two parallelograms, EA, AD being in a straight line. Let FG, drawn parallel to AC, meet BA produced in G. Then the triangle ABE equals the triangle ADG.

36. From AC, the diagonal of a square $ABCD$, cut off AE equal to one-fourth of AC, and join BE, DE. Shew that the figure $BADE$ is equal to twice the square on AE.

37. If ABC be a triangle, with the angles at B and C each double of the angle at A, prove that the square on AB is equal to the square on BC together with the rectangle AB, BC.

38. If two sides of a quadrilateral be parallel, the triangle contained by either of the other sides and the two straight lines drawn from its extremities to the middle point of the opposite side is half the quadrilateral.

39. Describe a parallelogram equal to and equiangular with a given parallelogram, and having a given altitude.

40. If the sides of a triangle taken in order be produced to twice their original lengths, and the outer extremities be joined, the triangle so formed will be seven times the original triangle.

41. If one of the acute angles of a right-angled isosceles triangle be bisected, the opposite side will be divided by the bisecting line into two parts, such that the square on one will be double of the square on the other.

42. ABC is a triangle, right-angled at B, and BD is drawn perpendicular to the base, and is produced to E until ECB is a right angle ; prove that the square on BC is equal to the sum of the rectangles AD, DC and BD, DE.

43. Shew that the sum of the squares on two unequal lines is greater than twice the rectangle contained by the lines.

44. From a given isosceles triangle cut off a trapezium, having the base of the triangle for one of its parallel sides, and having the other three sides equal.

45. If any number of parallelograms be constructed having their sides of given length, shew that the sum of the squares on the diagonals of each will be the same.

46. $ABCD$ is a right-angled parallelogram, and AB is double of BC; on AB an equilateral triangle is constructed: shew that its area will be less than that of the parallelogram.

47. A point O is taken within a triangle ABC, such that the angles BOC, COA, AOB are equal; prove that the squares on BC, CA, AB are together equal to the rectangles contained by OB, OC; OC, OA; OA, OB; and twice the sum of the squares on OA, OB, OC.

48. If the sides of an equilateral and equiangular hexagon be produced to meet, the angles formed by these lines are together equal to four right angles.

49. ABC is a triangle right-angled at A; in the hypotenuse two points D, E are taken such that $BD = BA$ and $CE = CA$; shew that the square on DE is equal to twice the rectangle contained by BE, CD.

50. Given one side of a rectangle which is equal in area to a given square, find the other side.

51. AB, AC are the two equal sides of an isosceles triangle; from B, BD is drawn perpendicular to AC, meeting it in D; shew that the square on BD is greater than the square on CD by twice the rectangle AD, CD.

BOOK III.

POSTULATE.

A POINT is within, or without, a circle, according as its distance from the centre is less, or greater than, the radius of the circle.

DEF. I. A straight line, as PQ, drawn so as to cut a circle $ABCD$, is called a SECANT.

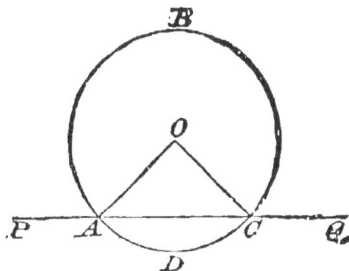

That such a line can only meet the circumference in *two* points may be shewn thus :

Some point within the circle is the centre ; let this be O. Join OA. Then (Ex. 1, I. 16) we can draw one, and only one, straight line from O, to meet the straight line PQ, such that it shall be equal to OA. Let this line be OC. Then A and C are the only points in PQ, which are *on* the circumference of the circle.

S. E. II.

D

DEF. II. The portion AC of the secant PQ, intercepted by the circle, is called a CHORD.

DEF. III. The two portions, into which a chord divides the circumference, as ABC and ADC, are called ARCS.

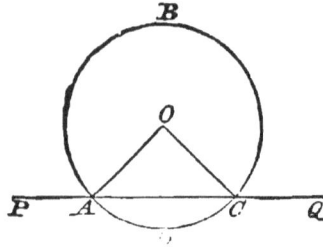

DEF. IV. The two figures into which a chord divides the circle, as ABC and ADC, that is, the figures, of which the boundaries are respectively the arc ABC and the chord AC, and the arc ADC and the chord AC, are called SEGMENTS of the circle.

DEF. V. The figure $AOCD$, whose boundaries are two radii and the arc intercepted by them, is called a SECTOR.

DEF. VI. A circle is said to be *described about* a rectilinear figure, when the circumference passes through each of the angular points of the figure.

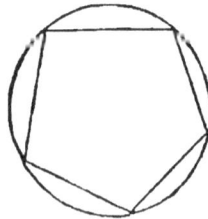

And the figure is said to be *inscribed* in the circle.

The line, which bisects a chord of a circle at right angles, must contain the centre.

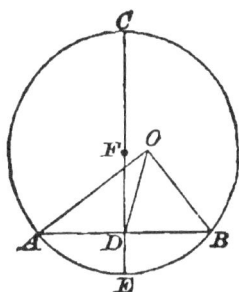

Let ABC be the given $\odot$.

Let the st. line CE bisect the chord AB at rt. angles in D.

Then the centre of the $\odot$ must lie in CE.

For if not, let O, a pt. out of CE, be the centre;
and join OA, OD, OB.

Then, in $\triangle$s ODA, ODB,

$\because AD = BD$, and DO is common, and $OA = OB$;

$\therefore \angle ODA = \angle ODB$;   I. c.

and $\therefore \angle ODB$ is a right $\angle$.   I. Def. 9

But $\angle CDB$ is a right $\angle$, by construction;

$\therefore \angle ODB = \angle CDB$, which is impossible;

$\therefore O$ is not the centre.

Thus it may be shewn that no point, out of CE, can be the centre, and $\therefore$ the centre must lie in CE.

COR. *If the chord CE be bisected in F, then F is the centre of the circle.*

PROPOSITION II. THEOREM.

*If any two points be taken in the circumference of a circle,
the straight line, which joins them, must fall within the
circle.*

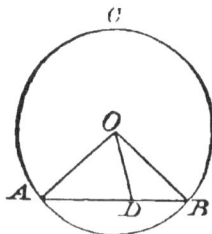

Let *A* and *B* be any two pts. in the O*ce* of the ⊙ *ABC.*

Then must the st. line AB fall within the ⊙.

Take any pt. *D* in the line *AB.*

Find *O* the centre of the ⊙.   III. 1, Cor.

Join *OA, OD, OB.*

Then ∵ ∠ *OAB* = ∠ *OBA*,   I. A.

and ∠ *ODB* is greater than ∠ *OAB*,   I. 16.

∴ ∠ *ODB* is greater than ∠ *OBA* ;

and ∴ *OB* is greater than *OD.*   I. 19.

∴ the distance of *D* from *O* is less than the radius of the ⊙ ,

and ∴ *D* lies within the ⊙.   Post.

And the same may be shown of any other pt. in *AB.*

∴ *AB* lies entirely within the ⊙.

Q. R. D.

PROPOSITION 111. THEOREM.

If a straight line, drawn through the centre of a circle, bisect a chord of the circle, which does not pass through the centre, it must cut it at right angles : and conversely, if it cut it at right angles, it must bisect it.

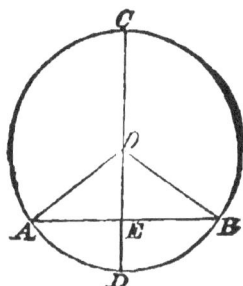

In the ⊙ *ABC,* let the chord *AB,* which does not pass through the centre *O,* be bisected in *E* by the diameter *CD.*

Then must CD be ⊥ *to AB.*

Join *OA, OB.*

Then in △s *AEO, BEO,*

∵ *AE=BE,* and *EO* is common, and *OA=OB,*

∴ ∠ *OEA =* ∠ *OEB.*   I. c.

Hence *OE* is ⊥ to *AB,*   I. Def. 9.

that is, *CD* is ⊥ to *AB.*

Next let *CD* be ⊥ to *AB.*

Then must CD bisect AB.

For ∵ *OA = OB,* and *OE* is common,

in the right-angled △s *AEO, BEO,*

∴ *AE=BE,*   I. E. Cor. p. 43.

that is, *CD* bisects *AB.*   Q. E. D.

Ex. 1. Shew that, if *CD* does not cut *AB* at right angles, it cannot bisect it.

Ex. 2. A line, which bisects two parallel chords in a circle, is also perpendicular to them.

Ex. 3. Through a given point within a circle, which is not the centre, draw a chord which shall be bisected in that point.

PROPOSITION IV. THEOREM.

If in a circle two chords, which do not both pass through the centre, cut one another, they do not bisect each other.

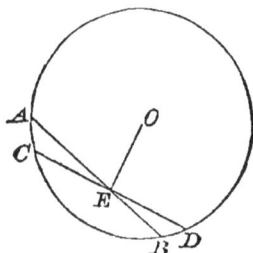

Let the chords AB, CD, which do not both pass through the centre, cut one another, in the pt. E, in the $\odot$ $ACBD$.

Then AB, CD do not bisect each other.

If one of them pass through the centre, it is plainly not bisected by the other, which does not pass through the centre.

But if neither pass through the centre, let, if it be possible, $AE = EB$ and $CE = ED$; find the centre O, and join OE.

Then ∵ OE, passing through the centre, bisects AB,

　　　　∴ ∠ OEA is a rt. ∠.　　　　　III. 3.

And ∵ OE, passing through the centre, bisects CD,

　　　　∴ ∠ OEC is a rt. ∠;　　　　　III. 3

∴ ∠ OEA = ∠ OEC, which is impossible ;

∴ AB, CD do not bisect each other.　　Q. E. D.

Ex. 1. Shew that the locus of the points of bisection of all parallel chords of a circle is a straight line.

Ex. 2. Shew that no parallelogram, except those which are rectangular, can be inscribed in a circle.

PROPOSITION V. THEOREM.

If two circles cut one another, they cannot have the same centre.

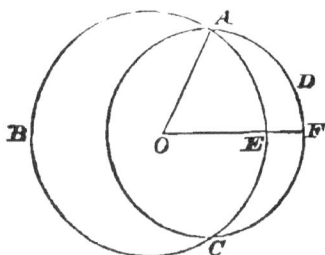

If it be possible, let O be the common centre of the ⊙s ABC, ADC, which cut one another in the pts. A and C.

Join OA, and draw OEF meeting the ⊙s in E and F.

Then ∵ O is the centre of ⊙ ABC,

$$\therefore OE = OA ; \qquad \text{I. Def. 13.}$$

and ∵ O is the centre of ⊙ ADC,

$$\therefore OF = OA ; \qquad \text{I. Def. 13.}$$

$\therefore OE = OF$, which is impossible ;

$\therefore O$ is not the common centre.

Q. E. D.

Ex. If two circles cut one another, shew that a line drawn through a point of intersection, terminated by the circumferences and parallel to the line joining the centres, is double of the line joining the centres.

Note. Circles which have the same centre are called *Concentric.*

NOTE 1. *On the Contact of Circles.*

DEF. VII. Circles are said to touch each other, which meet but do not cut each other.

One circle is said to touch another *internally,* when one point of the circumference of the former lies *on,* and no point *without,* the circumference of the other.

Hence for internal contact one circle must be smaller than the other.

Two circles are said to touch *externally,* when one point of the circumference of the one lies *on,* and no point *within* the circumference of the other.

N.B. No restriction is placed by these definitions on the number of points of contact, and it is not till we reach Prop. XIII. that we prove that there can be *but one point of contact.*

If one circle touch another internally, they cannot have the same centre.

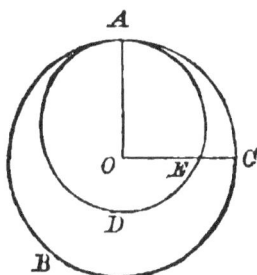

Let ⊙ *ADE* touch ⊙ *ABC* internally,

and let *A* be a point of contact.

Then *some* point *E* in the ○ce *ADE* lies *within* ⊙ *ABC*.

Def. 7.

If it be possible, let *O* be the common centre of the two ⊙s.

Join *OA*, and draw *OEC*, meeting the ○ces in *E* and *C*.

Then ∵ *O* is the the centre of ⊙ *ABC*,

∴ *OA* = *OC* ;   I. Def. 13.

and ∵ *O* is the centre of ⊙ *ADE*,

∴ *OA* = *OE*.   I. Def. 13.

Hence *OE* = *OC*, which is impossible ,

∴ *O* is not the common centre of the two ⊙s.

Q. E. D.

PROPOSITION VII. THEOREM.

If from any point within a circle, which is not the centre, straight lines be drawn to the circumference, the greatest of these lines is that which passes through the centre.

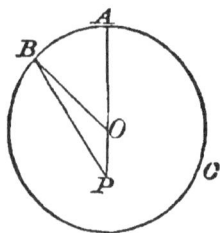

Let ABC be a $\odot$, of which O is the centre.

From P, any pt. within the $\odot$, draw the st. line PA, passing through O and meeting the $\odot$ce in A.

Then must PA be greater than any other st. line, drawn from P to the $\odot$ce.

For let PB be any other st. line, drawn from P to meet the $\odot$ce in B, and join BO.

Then ∵ $AO = BO$,

∴ $AP =$ sum of BO and OP.

But the sum of BO and OP is greater than BP,   I. 20.

and ∴ AP is greater than BP.   Q. E. D.

Ex. 1. If AP be produced to meet the circumference in D, shew that PD is less than any other straight line that can be drawn from P to the circumference.

Ex. 2. Shew that PB continually decreases, as B passes from A to D.

Ex. 3. Shew that two straight lines, but not three, that shall be equal, can be drawn from P to the circumference.

PROPOSITION VIII. THEOREM.

If from any point without a circle straight lines be drawn to the circumference, the least of these lines is that which, when produced, passes through the centre, and the greatest is that which passes through the centre.

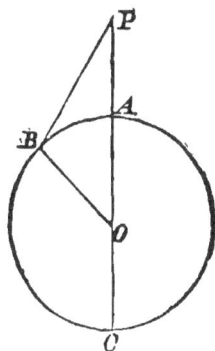

Let ABC be a ⊙, of which O is the centre.

From P any pt. outside the ⊙, draw the st. line $PAOC$, meeting the ⊙ce in A and C.

Then must PA be less, and PC greater, than any other st. line drawn from P to the ⊙ce.

For let PB be any other st. line drawn from P to meet the ⊙ce in B, and join BO.

Then ∵ sum of PB and BO is greater than OP, I. 20.

∴ sum of PB and BO is greater than sum of AP and AO.

But $BO = AO$;

∴ PB is greater than AP.

Again ∵ PB is less than the sum of PO, OB, I. 20.

∴ PB is less than the sum of PO, OC ;

∴ PB is less than PC. Q. E. D.

Ex. 1. Shew that PB continually increases as B passes from A to C.

Ex. 2. Shew that from P two straight lines, but not three, that shall be equal, can be drawn to the circumference.

NOTE. From Props. VII. and VIII. we deduce the following Corollary, which we shall use in the proof of Props. XI. and XIII.

COR. *If a point be taken, within or without a circle, of all straight lines drawn from it to the circumference, the greatest is that which meets the circumference after passing through the centre.*

Proposition IX. Theorem.

If a point be taken within a circle, from which there fall more than two equal straight lines to the circumference, that point is the centre of the circle.

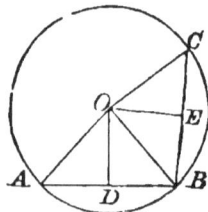

Let O be a pt. in the ⊙ ABC from which more than two st. lines OA, OB, OC, drawn to the ⊙ce, are equal.

Then must O be the centre of the ⊙.

Join AB, BC, and draw OD, OE ⊥ to AB, BC.

Then ∵ $OA = OB$, and OD is common,

in the right-angled △ s AOD, BOD,

∴ $AD = DB$;   I. E. Cor. p. 43.

∴ the centre of the ⊙ is in DO.   III. 1.

Similarly it may be shown that

the centre of the ⊙ is EO ;

∴ O is the centre of the ⊙.

Q. E. D.

PROPOSITION X. THEOREM.

Two circles cannot have more than two points common to both, without coinciding entirely.

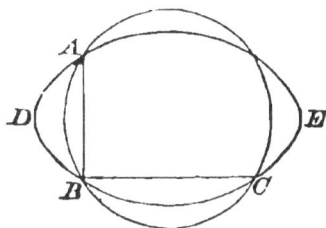

If it be possible, let ABC and ADE be two ⊙s which have more than two pts. in common, as A, B, C.

Join AB, BC.

Then ∵ AB is a chord of each circle,

∴ the centre of each circle lies in the straight line, which bisects AB at right angles ;   III. 1.

and ∵ BC is a chord of each circle,

∴ the centre of each circle lies in the straight line, which bisects BC at right angles.   III. 1.

∴ the centre of each circle is the point, in which the two straight lines, which bisect AB and BC at right angles, meet.

∴ the ⊙s ABC, ADE have a common centre, which is impossible ;   III. 5 and 6.

∴ two ⊙s cannot have more than two pts. common to both.

Q. E. D.

NOTE. We here insert two Propositions, Eucl. III. 25 and IV. 5, which are closely connected with Theorems I. and X. of this book. The learner should compare with this portion of the subject the note on Loci, p. 103.

PROPOSITION A.   PROBLEM.   (Eucl. III. 25.)

An arc of a circle being given, to complete the circle of which it is a part.

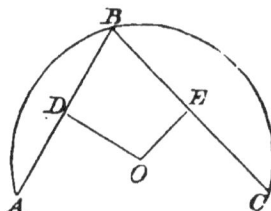

Let *ABC* be the given arc.

It is required to complete the ⊙ of which ABC is a part.

Take *B*, any pt. in arc *ABC*, and join *AB, BC*.

From *D* and *E*, the middle pts. of *AB* and *BC*,

    draw *DO, EO*, ⊥s to *AB, BC*, meeting in *O*.

Then ∵ *AB* is to be a chord of the ⊙,

    ∴ centre of the ⊙ lies in *DO* ;   III. 1.

and ∵ *BC* is to be a chord of the ⊙,

    ∴ centre of the ⊙ lies in *EO*.   III. 1.

Hence *O* is the centre of the ⊙ of which *ABC* is an arc, and if a ⊙ be described, with centre *O* and radius *OA*, this will be the ⊙ required.

Q. E. F.

PROPOSITION B. PROBLEM. (Eucl. IV. 5.)

To describe a circle about a given triangle.

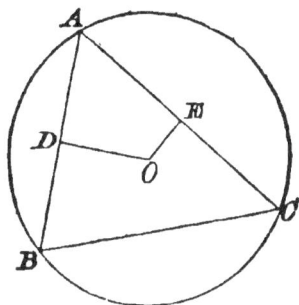

Let ABC be the given $\triangle$.

It is required to describe a $\odot$ about the $\triangle$.

From D and E, the middle pts. of AB and AC, draw DO, EO, ⊥ s to AB, AC, and let them meet in O.

Then ∵ AB is to be a chord of the $\odot$,

∴ centre of the $\odot$ lies in DO.   III. 1.

And ∵ AC is to be a chord of the $\odot$,

∴ centre of the $\odot$ lies in EO.   III. 1.

Hence O is the centre of the $\odot$ which can be described about the $\triangle$, and if a $\odot$ be described with centre O and radius OA, this will be the $\odot$ required.

Q. E. F.

Ex. If BAC be a right angle, show that O will coincide with the middle point of BC.

Proposition XI. Theorem.

If one circle touch another internally at any point, the centre of the interior circle must lie in that radius of the other circle which passes through that point of contact.

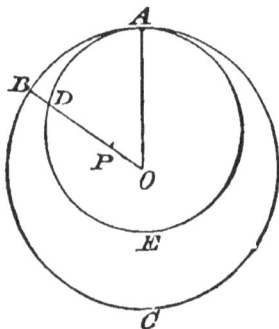

Let the ⊙ ADE touch the ⊙ ABC internally, and let A be a pt. of contact.

Find O the centre of ⊙ ABC, and join OA.

Then must the centre of ⊙ ADE lie in the radius OA.

For if not, let P be the centre of ⊙ ADE.

Join OP, and produce it to meet the ⊙ces in D and B.

Then ∵ P is the centre of ⊙ ADE, and from O are drawn to the ⊙ce of ADE the st. lines OA, OD, of which OD passes through P,

∴ OD is greater than OA.   III. 8, Cor.

But $OA = OB$;

∴ OD is greater than OB,

which is impossible.

∴ the centre of ⊙ ADE is not out of the radius OA.

∴ it lies in OA.

Q. E. D.

If two circles touch one another externally at any point, the straight line joining the centre of one with that point of contact must when produced pass through the centre of the other.

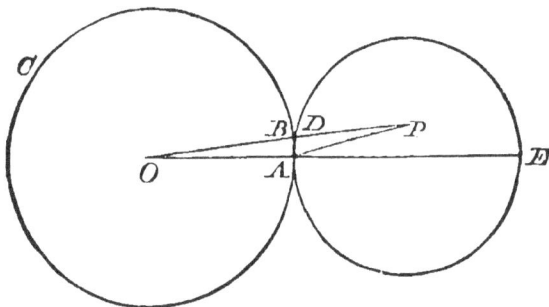

Let ⊙ *ABC* touch ⊙ *ADE* externally at the pt. *A*.

Let *O* be the centre of ⊙ *ABC*.

Join *OA*, and produce it to *E*.

Then must the centre of ⊙ ADE lie in AE.

For if not, let *P* be the centre of ⊙ *ADE*.

Join *OP* meeting the ⊙s in *B*, *D*; and join *AP*.

Then ∵ *OB* = *OA*,

and *PD* = *AP*,

∴ *OB* and *PD* together = *OA* and *AP* together;

∴ *OP* is not less than *OA* and *AP* together.

But *OP* is less than *OA* and *AP* together,   I. 20.

which is impossible;

∴ the centre of ⊙ *ADE* cannot lie out of *AE*.

Q. E. D.

Ex. Three circles touch one another externally, whose centres are *A*, *B*, *C*. Shew that the difference between *AB* and *AC* is half as great as the difference between the diameters of the circles, whose centres are *B* and *C*.

One circle cannot touch another at more points than one, whether it touch it internally or externally.

First let the ⊙ *ADE* touch the ⊙ *ABC* internally at pt. *A*.

Then there can be no other point of contact.

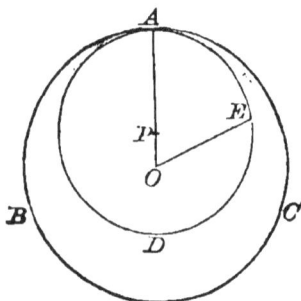

Take *O* the centre of ⊙ *ABC*

Then *P*, the centre of ⊙ *ADE*, lies in *OA*.   III. 11.

Take any pt. *E* in the ◯ce of the ⊙ *ADE*, and join *OE*.

Then ∵ from *O*, a pt. within or without the ⊙ *ADE*, two lines *OA*, *OE* are drawn to the ◯ce, of which *OA* passes through the centre *P*,

∴ *OA* is greater than *OE*,   III. 8, Cor.

and ∴ *E* is a point *within* the ⊙ *ABC*.   Post.

Similarly it may be shewn that every pt. of the ◯ce of the ⊙ *ADE*, except *A*, lies *within* the ⊙ *ABC* ;

∴ *A* is the only point at which the ⊙s meet.

Next. let the ⊙s ABC, ADE touch *externally* at the pt. A.

Then there can be no other point of contact.

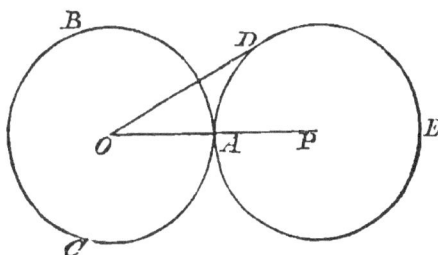

Take O the centre of the ⊙ ABC.

Then P, the centre of the ⊙ ADE, lies in OA produced.

III. 12.

Take any pt. D in the ◯ce of the ⊙ ADE, and join OD.

Then ∵ from O, a pt. without the ⊙ ADE, two lines OA, OD are drawn to the ◯ce, of which OA when produced passes through the centre P,

∴ OD is greater than OA ;   III. 8.

∴ D is a point *without* the ⊙ ABC.   Post.

Similarly, it may be shewn that every pt. of the ◯ce of ADE, except A, lies *without* the ⊙ ABC ;

∴ A is the only point at which the ⊙s meet.

Q. E. D.

DEF. VIII. The DISTANCE of a chord from the centre is measured by the length of the perpendicular drawn from the centre to the chord.

PROPOSITION XIV.   THEOREM.

Equal chords in a circle are equally distant from the centre; and conversely, those which are equally distant from the centre, are equal to one another.

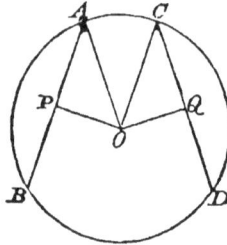

Let the chords AB, CD in the $\odot$ $ABDC$ be equal.

Then must AB and CD be equally distant from the centre O.

Draw OP and OQ $\perp$ to AB and CD; and join AO, CO.

Then P and Q are the middle pts. of AB and CD: III. 3.

and $\because$ $AB = CD$, $\therefore$ $AP = CQ$.

Then $\because$ $AP = CQ$, and $AO = CO$,

in the right-angled $\triangle$s AOP, COQ,

$\therefore$ $OP = OQ$;   I. E. Cor. p. 43.

and $\therefore$ AB and CD are equally distant from O.   Def. 8.

Next, let AB and CD be equally distant from O.

Then must $AB = CD$.

For $\because$ $OP = OQ$, and $AO = CO$,

in the right-angled $\triangle$s AOP, COQ,

$\therefore$ $AP = CQ$,   I. E. Cor.

and $\therefore$ $AB = CD$.

Q. E. D.

Ex. In a circle, whose diameter is 10 inches, a chord is drawn, which is 8 inches long. If another chord be drawn, at a distance of 3 inches from the centre, shew whether it is equal or not to the former.

PROPOSITION XV.   THEOREM.

The diameter is the greatest chord in a circle, and of all others that which is nearer to the centre is always greater than one more remote ; and the greater is nearer to the centre than the less.

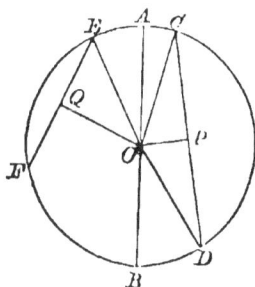

Let AB be a diameter of the $\odot$ $ABDC$, whose centre is O, and let CD be any other chord, not a diameter, in the $\odot$, nearer to the centre than the chord EF.

Then must AB be greater than CD, and CD greater than EF.

Draw OP, OQ $\perp$ to CD and EF ; and join OC, OD, OE.

Then $\because$ $AO = CO$, and $OB = OD$,   I. Def. 13.

$\therefore$ $AB =$ sum of CO and OD,

and $\therefore$ AB is greater than CD.   I. 20.

Again, $\because$ CD is nearer to the centre than EF,

$\therefore$ OP is less than OQ.   Def. 8.

Now $\because$ sq. on $OC =$ sq. on OE,

$\therefore$ sum of sqq. on OP, $PC =$ sum of sqq. on OQ, QE.   I. 47.

But sq. on OP is less than sq. on OQ ;

$\therefore$ sq. on PC is greater than sq. on QE ;

$\therefore$ PC is greater than QE ;

and $\therefore$ CD is greater than EF.

Next, let CD be greater than EF.

Then must CD be nearer to the centre than EF.

For $\because$ CD is greater than EF,

$\therefore$ PC is greater than QE.

Now the sum of sqq. on OP, PC=sum of sqq. on OQ, Q ".
But sq. on PC is greater than sq. on QE ;
∴ sq. on OP is less than sq. on OQ ;
∴ OP is less than OQ ;
and ∴ CD is nearer to the centre than EF.

Q. E. D.

Ex. 1. Draw a chord of given length in a given circle, which shall be bisected by a given chord.

Ex. 2. If two isosceles triangles be of equal altitude, and the sides of one be equal to the sides of the other, shew that their bases must be equal.

Ex. 3. Any two chords of a circle, which cut a diameter in the same point and at equal angles, are equal to one another.

DEF. IX. *A straight line is said to be a* TANGENT *to, or to touch, a circle, when it meets and, being produced, does not cut the circle.*

From this definition it follows that the tangent meets the circle in one point only, for if it met the circle in two points it would cut the circle, since the line joining two points in the circumference is, being produced, a secant. (III. 2.)

DEF. X. If from any point in a circle a line be drawn at right angles to the tangent at that point, the line is called a NORMAL to the circle at that point.

DEF. XI. A rectilinear figure is said to be *described about* a circle, when each side of the figure touches the circle.

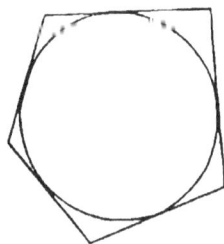

And the circle is said to be *inscribed* in the figure.

The straight line drawn at right angles to the diameter of a circle, from the extremity of it, is a tangent to the circle.

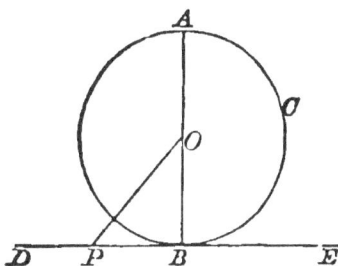

Let ABC be a $\odot$, of which the centre is O, and the diameter AOB.

Through B draw DE at right angles to AOB.   I. 11.

Then must DE be a tangent to the $\odot$.

Take any point P in DE, and join OP.

Then, $\because \angle OBP$ is a right angle,

$\therefore \angle OPB$ is less than a right angle,   I. 17.

and $\therefore$ OP is greater than OB.   I. 19.

Hence P is a point without the $\odot$ ABC.   Post.

In the same way it may be shewn that every point in DE, or DE produced in either direction, except the point B, lies without the $\odot$;

$\therefore DE$ is a tangent to the $\odot$.   Def. 9.

Q. E. D.

PROPOSITION XVII. PROBLEM.

To draw a straight line from a given point, either WITHOUT
or ON *the circumference, which shall touch a given circle.*

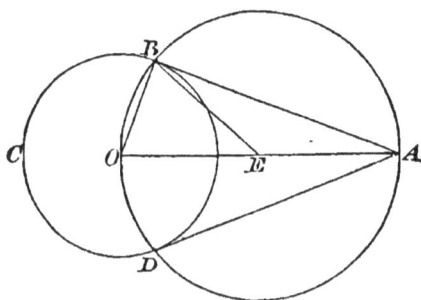

Let A be the given pt., *without* the ⊙ BCD.
Take O the centre of ⊙ BCD, and join OA.
Bisect OA in E, and with centre E and radius EO describe
⊙ $ABOD$, cutting the given ⊙ in B and D.
Join AB, AD.   *These are tangents to the* ⊙ BCD.
Join BO, BE.

Then ∵ $OE = BE$, ∴ ∠ $OBE = $ ∠ BOE ;   I. A.
∴ ∠ AEB = twice ∠ OBE ;   I. 32.
and ∵ $AE = BE$, ∴ ∠ $ABE = $ ∠ BAE ;   I. A.
∴ ∠ OEB = twice ∠ ABE ;   I. 32.
∴ sum of ∠s AEB, OEB = twice sum of ∠s OBE, ABE,
that is, two right angles = twice ∠ OBA ;
∴ ∠ OBA is a right angle,
and ∴ AB is a tangent to the ⊙ BCD.   III. 16.
Similarly it may be shewn that AD is a tangent to ⊙ BCD.
Next, let the given pt. be on the ⊙ce of the ⊙, as B.
Then, if BA be drawn ⊥ to the radius OB,
BA is a tangent to the ⊙ at B.   III. 16.

Q. E. D.

Ex. 1. Shew that the two tangents, drawn from a point without the circumference to a circle, are equal.

Ex. 2. If a quadrilateral $ABCD$ be described about a circle,
shew that the sum of AB and CD is equal to the sum of AD
and BC.

If a straight line touch a circle, the straight line drawn from the centre to the point of contact must be perpendicular to the line touching the circle.

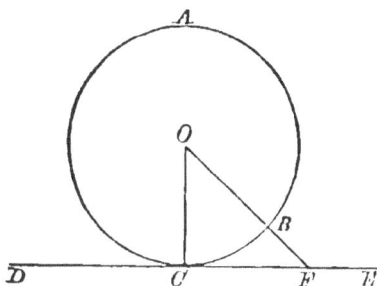

Let the st. line DE touch the $\odot$ ABC in the pt. C.

Find O the centre, and join OC.

Then must OC be $\perp$ to DE.

For if it be not, draw $OBF \perp$ to DE, meeting the $\odot$ce in B.

Then $\because \angle OFC$ is a rt. angle,

$\therefore \angle OCF$ is less than a rt. angle,   I. 17.

and $\therefore OC$ is greater than OF.   I. 19.

But $OC = OB$,

$\therefore OB$ is greater than OF, which is impossible;

$\therefore OF$ is not $\perp$ to DE, and in the same way it may be shewn that no other line drawn from O, but OC, is $\perp$ to DE;

$\therefore OC$ is $\perp$ to DE.

Q. E. D.

Ex. If two straight lines intersect, the centres of all circles touched by both lines lie in two lines at right angles to each other.

NOTE. Prop. XVIII. might be stated thus :—*All radii of a circle are normals to the circle at the points where they meet the circumference.*

PROPOSITION XIX.   THEOREM.

If a straight line touch a circle, and from the point of con-
tact a straight line be drawn at right angles to the touching line,
the centre of the circle must be in that line.

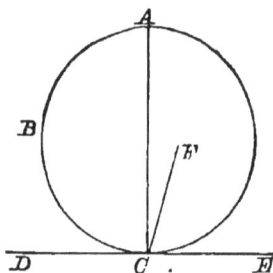

Let the st. line DE touch the $\odot$ ABC at the pt. C, and
from C let CA be drawn $\perp$ to DE.

Then must the centre of the $\odot$ be in CA.

For if not, let F be the centre, and join FC.

Then $\because$ DCE touches the $\odot$, and FC is drawn from centre
to pt. of contact,

$\therefore \angle FCE$ is a rt. angle.   III. 18.

But $\angle ACE$ is a rt. angle.

$\therefore \angle FCE = \angle ACE$, which is impossible.

In the same way it may be shewn that no pt. out of CA
can be the centre of the $\odot$;

$\therefore$ the centre of the $\odot$ lies in CA.

Q. E. D.

Ex. Two concentric circles being described, if a chord of
the greater touch the less, the parts of the chord, intercepted
between the two circles, are equal.

NOTE.   Prop. XIX. might be stated thus :— *Every normal to*
a circle passes through the centre.

The angle at the centre of a circle is double of the angle at the circumference, subtended by the same arc.

Let ABC be a $\odot$, O the centre,

BC any arc, A any pt. in the $\bigcirc$ce.

Then must $\angle BOC = $ twice $\angle BAC$.

First, suppose O to be in one of the lines containing the $\angle BAC$.

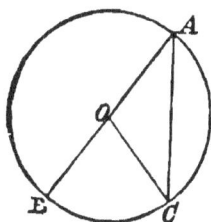

Then $\because OA = OC$,

$\therefore \angle OCA = \angle OAC$;   I. A.

$\therefore$ sum of $\angle$ s OCA, $OAC = $ twice $\angle OAC$.

But $\angle BOC = $ sum of $\angle$ s OCA, OAC,   I. 32.

$\therefore \angle BOC = $ twice $\angle OAC$.

that is, $\angle BOC = $ twice $\angle BAC$.

Next, suppose O to be within (fig 1), or without (fig. 2) the
$\angle BAC$.

Fig. 1.   Fig. 2.

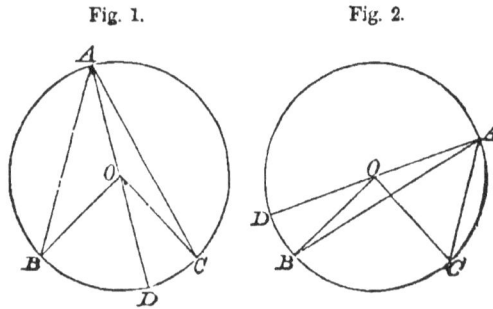

Join AO, and produce it to meet the ⊙ce in D.

Then, as in the first case,

$$\angle COD = \text{twice} \angle CAD,$$

$$\text{and} \angle BOD = \text{twice} \angle BAD\ ;$$

∴ fig. 1, sum of $\angle$ s COD, BOD = twice sum of $\angle$ s CAD,
BAD,

that is, $\angle BOC = \text{twice} \angle BAC.$

And, fig. 2, difference of $\angle$ s COD, BOD = twice difference
of $\angle$ s CAD, BAD, that is, $\angle BOC = \text{twice} \angle BAC.$

Q. E. D.

Ex. From any point in a straight line, touching a circle,
a straight line is drawn through the centre, and is terminated
by the circumference ; the angle between these two straight
lines is bisected by a straight line, which intersects the straight
line joining their extremities. Shew that the angle between
the last two lines is half a right angle

NOTE 2. *On Flat and Reflex Angles.*

We have already explained (Note 3, Book I., p. 28) how Euclid's definition of an angle may be extended with advantage, so as to include the conception of an angle equal to two right angles : and we now proceed to shew how the Definition given in that Note may be extended, so as to embrace angles greater than two right angles.

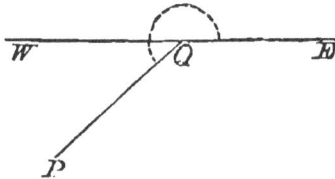

Let WQ be a straight line, and QE its continuation.

Then, by the Definition, the angle made by WQ and QE, which we propose to call a FLAT ANGLE, is equal to two right angles.

Now suppose QP to be a straight line, which revolves about the fixed point Q, and which at first coincides with QE.

When QP, revolving from right to left, coincides with QW, it has described an angle equal to two right angles.

When QP has continued its revolution, so as to come into the position indicated in the diagram, it has described an angle EQP, indicated by the dotted line, greater than two right angles, and this we call a REFLEX ANGLE.

To assist the learner, we shall mark these angles with dotted lines in the diagrams.

Admitting the existence of angles, equal to and greater than two right angles, the Proposition last proved may be extended, as we now proceed to shew.

PROPOSITION C. THEOREM.

The angle, not less than two right angles, at the centre of a circle is double of the angle at the circumference, subtended by the same arc.

Fig. 1.   Fig. 2.

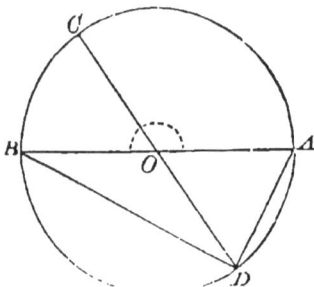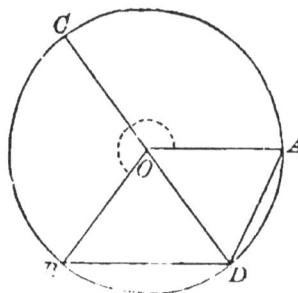

In the ⊙ *ACBD*, let the angles *AOB* (not less than two right angles) at the centre, and *ADB* at the circumference, be subtended by the same arc *ACB*.

Then must ∠ AOB = twice ∠ ADB.

Join *DO*, and produce it to meet the arc *ACB* in *C*.

Then ∵ ∠ *AOC* = twice ∠ *ADO*,   III. 20.

and ∠ *BOC* = twice ∠ *BDO*,   III. 20.

∴ sum of ∠ s *AOC*, *BOC* = twice sum of ∠ s *ADO*, *BDO*,

that is, ∠ *AOB* = twice ∠ *ADB*.

Q. E. D.

NOTE. In fig. 1, ∠ *AOB* is drawn a flat angle,

and in fig. 2, ∠ *AOB* is drawn a reflex angle.

DEF. XII. The angle in a segment is the angle contained by two straight lines drawn from any point in the arc to the extremities of the chord.

PROPOSITION XXI. THEOREM.

The angles in the same segment of a circle are equal to one another.

Fig. 1.   Fig. 2.

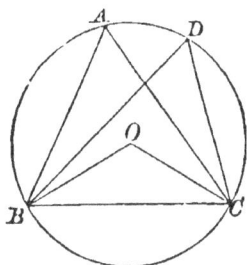 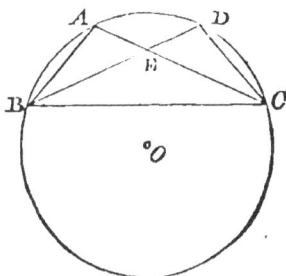

Let BAC, BDC be angles in the same segment $BADC$.
      Then must ∠ BAC = ∠ BDC.
First, when segment $BADC$ is greater than a semicircle,
      From O, the centre, draw OB, OC.   (Fig. 1.)
      Then, ∵ ∠ BOC = twice ∠ BAC,   III. 20.
       and ∠ BOC = twice ∠ BDC,   III. 20.
      ∴ ∠ BAC = ∠ BDC.
Next, when segment $BADC$ is less than a semicircle,
      Let E be the pt. of intersection of AC, DB. (Fig. 2.)
      Then ∵ ∠ ABE = ∠ DCE, by the first case,
        and ∠ BEA = ∠ CED,   I. 15.
      ∴ ∠ EAB = ∠ EDC,   I. 32.
      that is, ∠ BAC = ∠ BDC.   Q. E. D.

Ex. 1. Shew that, by assuming the possibility of an angle being greater than two right angles, both the cases of this proposition may be included in one.

Ex. 2. If two straight lines, whose extremities are in the circumference of a circle, cut one another, the triangles formed by joining their extremities are equiangular to each other.

PROPOSITION XXII. THEOREM.

The opposite angles of any quadrilateral figure, inscribed in a circle, are together equal to two right angles.

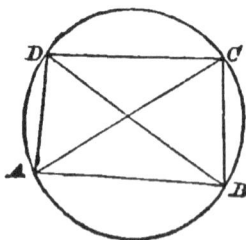

Let $ABCD$ be a quadrilateral fig. inscribed in a ⊙.

Then must each pair of its opposite ∠ s be together equal to two rt. ∠ s.

Draw the diagonals AC, BD.

Then ∵ ∠ ADB = ∠ ACB, in the same segment,   III. 21.

      and ∠ BDC = ∠ BAC, in the same segment ;   III. 21.

∴ sum of ∠ s ADB, BDC = sum of ∠ s ACB, BAC ;

      that is, ∠ ADC = sum of ∠ s ACB, BAC.

            Add to each ∠ ABC.

Then ∠ s ADC, ABC together = sum of ∠ s ACB, BAC, ABC ;

      and ∴ ∠ s ADC, ABC together = two right ∠ s.   I. 32.

Similarly, it may be shewn,

      that ∠ s BAD, BCD together = two right ∠ s.

                                                  Q. E. D.

NOTE.—Another method of proving this proposition is given on page 177.

Ex. 1. If one side of a quadrilateral figure inscribed in a circle be produced, the exterior angle is equal to the opposite angle of the quadrilateral.

Ex. 2. If the sides *AB*, *DC* of a quadrilateral inscribed in a circle be produced to meet in *E*, then the triangles *EBC*, *EAD* will be equiangular.

Ex. 3. Shew that a circle cannot be described about a rhombus.

Ex. 4. The lines, bisecting any angle of a quadrilateral figure inscribed in a circle and the opposite exterior angle, meet in the circumference of the circle.

Ex. 5. *AB*, a chord of a circle, is the base of an isosceles triangle, whose vertex *C* is without the circle, and whose equal sides meet the circle in *D*, *E* : shew that *CD* is equal to *CE*.

Ex. 6. If in any quadrilateral the opposite angles be to-gether equal to two right angles, a circle may be described about that quadrilateral.

Propositions XXIII. and XXIV., not being required in the method adopted for proving the subsequent Propositions in this book, are removed to the Appendix. Proposition XXV. has been already proved.

NOTE 3. *On the Method of Superposition, as applied to Circles.*

In Props. XXVI. XXVII. XXVIII. XXIX. we prove certain relations existing between chords, arcs, and angles in equal circles. As we shall employ the Method of Superposition, we must state the principles which render this method appli-cable, as a test of equality, in the case of figures with circular boundaries.

DEF. XIII. *Equal circles are those, of which the radii are equal.*

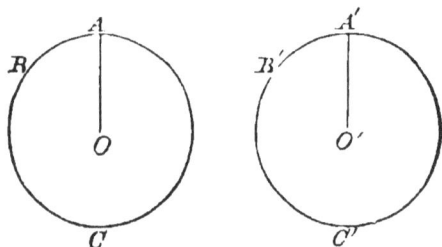

For suppose ABC, $A'B'C'$ to be circles, of which the radii are equal.

Then if $\odot$ $A'B'C'$ be applied to $\odot$ ABC, so that O', the centre of $A'B'C'$, coincides with O, the centre of ABC, it is evident that any *particular* point A' in the $\bigcirc$ce of the former must coincide with *some* point A in $\bigcirc$ce of the latter, because of the equality of the radii $O'A'$ and OA.

Hence $\bigcirc$ce $A'B'C'$ must coincide with $\bigcirc$ce ABC,

that is, $\odot$ $A'B'C' = \odot ABC$.

Further, when we have applied the circle $A'B'C'$ to the circle ABC, so that the centres coincide, we may imagine ABC to remain fixed, while $A'B'C'$ revolves round the common centre. Hence we may suppose any particular point B' in the circumference of $A'B'C'$ to be made to coincide with any particular point B in the circumference of ABC.

Again, any radius $O'A'$ of the circle $A'B'C'$ may be made to coincide with any radius OA of the circle ABC.

Also, if $A'B'$ and AB be equal arcs, they may be made to coincide.

Again, every diameter of a circle divides the circle into equal segments.

For let AOB be a diameter of the circle $ACBD$, of which O is the centre. Suppose the segment ACB to be applied to the segment ADB, so as to keep AB a common boundary : then the arc ACB must coincide with the arc ADB, because every point in each is equally distant from O.

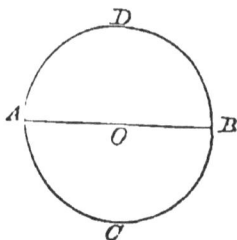

In equal circles, the arcs, which subtend equal angles, whether they be at the centres or at the circumferences, must be equal.

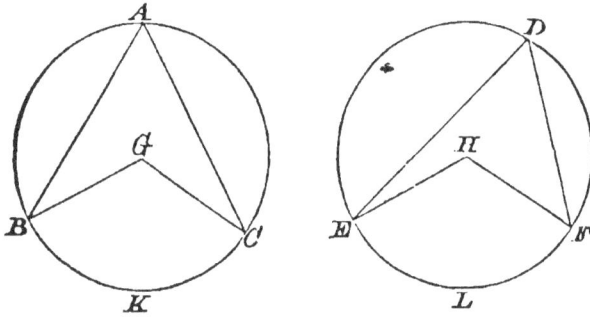

Let *ABC, DEF* be equal circles, and let ∠ s *BGC, EHF* at their centres, and ∠ s *BAC, EDF* at their ○ces, be equal.

Then must arc BKC=arc ELF.

For, if ⊙ *ABC* be applied to ⊙ *DEF*,

so that *G* coincides with *H*, and *GB* falls on *HE*,

then, ∵ *GB=HE*, ∴ *B* will coincide with *E*.

And ∵ ∠ *BGC=* ∠ *EHF*, ∴ *GC* will fall on *HF*;

and ∵ *GC=HF*, ∴ *C* will coincide with *F*.

Then ∵ *B* coincides with *E* and *C* with *F*,

∴ arc *BKC* will coincide with and be equal to arc *ELF*.

Q. E. D.

Cor. Sector *BGCK* is equal to sector *EHFL*.

Note. This and the three following Propositions are, and will hereafter be assumed to be, true for *the same circle* as well as for *equal circles.*

PROPOSITION XXVII. THEOREM.

In equal circles, the angles, which are subtended by equal arcs, whether they are at the centres or at the circumferences, must be equal.

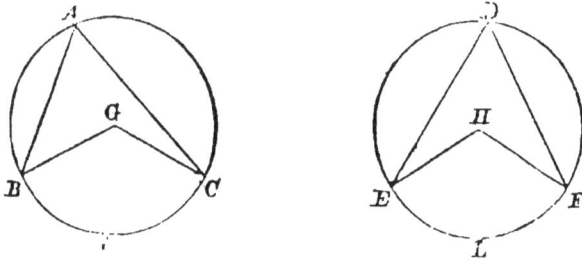

Let ABC, DEF be equal circles, and let ∠ s BGC, EHF at their centres, and ∠ s BAC, EDF at their ⊙ces, be subtended by equal arcs BKC, ELF.

Then must ∠ BGC= ∠ EHF, and ∠ BAC= ∠ EDF.

For, if ⊙ ABC be applied to ⊙ DEF,
so that G coincides with H, and GB falls on HE,
then ∵ $GB=HE$, ∴ B will coincide with E ;
and ∵ arc BKC=arc ELF, ∴ C will coincide with F.
Hence, GC will coincide with HF.
Then ∵ BG coincides with EH, and GC with HF,
∴ ∠ BGC will coincide with and be equal to ∠ EHF.
Again, ∵ ∠ BAC=half of ∠ BGC,   III. 20.
and ∠ EDF=half of ∠ EHF,   III. 20.
∴ ∠ BAC= ∠ EDF.   I. Ax. 7.

Q. E. D,

Ex. 1. If, in a circle, AB, CD be two arcs of given magnitude, and AC, BD be joined to meet in E, shew that the angle AEB is invariable.

Ex. 2. The straight lines joining the extremities of the chords of two equal arcs of the same circle, towards the same parts, are parallel to each other.

In equal circles, the arcs, which are subtended by equal chords, must be equal, the greater to the greater, and the less to the less.

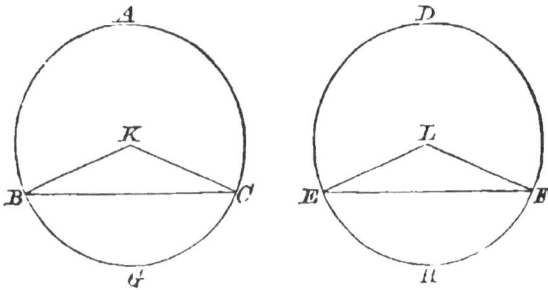

Let ABC, DEF be equal circles, and BC, EF equal chords, subtending the major arcs BAC, EDF,
and the minor arcs BGC, EHF.

Then must arc $BAC = $ arc EDF, and arc $BGC = $ arc EHF.

Take the centres K, L, and join KB, KC, LE, LF.

Then ∵ $KB=LE$, and $KC=LF$, and $BC=EF$,

∴ ∠ $BKC = $ ∠ ELF.   I. C.

Hence, if ⊙ ABC be applied to ⊙ DEF,

so that K coincides with L, and KB falls on LE,

then ∵ ∠ $BKC = $ ∠ ELF, ∴ KC will fall on LF;

and ∵ $KC = LF$, ∴ C will coincide with F.

Then ∵ B coincides with E, and C with F,

∴ arc BAC will coincide with and be equal to arc EDF,

and arc BGC...EHF.

Q. E. D.

Ex. 1. If, in a circle $ABCD$, the chord AB be equal to the chord DC, AD must be parallel to BC.

Ex. 2. If a straight line, drawn from A the middle point of an arc BC, touch the circle, shew that it is parallel to the chord BC.

Ex. 3. If two equal chords, in a given circle, cut one another, the segments of the one shall be equal to the segments of the other, each to each.

PROPOSITION XXIX. THEOREM.

In equal circles, the chords, which subtend equal arcs, must be equal.

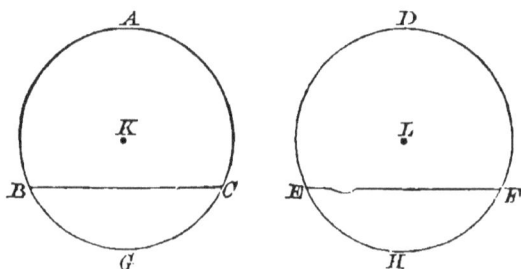

Let *ABC*, *DEF* be equal circles, and let *BC*, *EF* be chords subtending the equal arcs *BGC*, *EHF*.

Then must chord BC = chord EF.

Take the centres *K*, *L*.

Then, if ⊙ *ABC* be applied to ⊙ *DEF*,

so that *K* coincides with *L*, and *B* with *E*,

and arc *BGC* falls on arc *EHF*,

∵ arc *BGC*=arc *EHF*, ∴ *C* will coincide with *F*.

Then ∵ *B* coincides with *E* and *C* with *F*,

∴ chord *BC* must coincide with and be equal to chord *EF*,

Q. E. D.

Ex. 1. The two straight lines in a circle, which join the extremities of two parallel chords, are equal to one another.

Ex. 2. If three equal chords of a circle, cut one another in the same point within the circle, that point is the centre.

NOTE 4. *On the Symmetrical properties of the Circle with regard to its diameter.*

The brief remarks on Symmetry in pp. 107, 108 may now be extended in the following way :

A *figure* is said to be symmetrical with regard to a line, when every perpendicular to the line meets the figure at points which are equidistant from the line.

Hence a Circle is Symmetrical with regard to its Diameter, because the diameter *bisects* every chord, to which it is perpendicular.

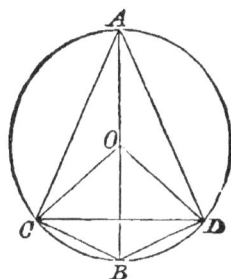

Further, suppose *AB* to be a diameter of the circle *ACBD*, of which *O* is the centre, and *CD* to be a chord perpendicular to *AB*.

Then, if lines be drawn as in the diagram, we know that *AB* bisects

    (1.) The chord *CD*,   III. 1.

    (2.) The arcs *CAD* and *CBD*,   III. 26.

    (3.) The angles *CAD*, *COD*, *CBD*, and the reflex angle *DOC*.   I. 4.

        Also, chord *CB* =chord *DB*,   I. 4.

        and chord *AC*=chord *AD*.   I. 4.

These Symmetrical relations should be carefully observed, because they are often suggestive of methods for the solution of problems.

Proposition XXX. Problem.

To bisect a given arc.

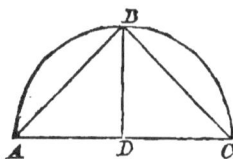

Let ABC be the given arc.

It is required to bisect the arc ABC.

Join AC, and bisect the chord AC in D.   I. 10.

From D draw $DB \perp$ to AC.   I. 11.

Then will the arc ABC be bisected in B.

Join BA, BC.

Then, in $\triangle$s ADB, CDB,

∵ $AD = CD$, and DB is common, and $\angle ADB = \angle CDB$,

∴ $BA = BC$.   I. 4.

But, in the same circle, the arcs, which are subtended by equal chords, are equal, the greater to the greater and the less to the less ;   III. 28.

and ∵ BD, if produced, is a diameter,

∴ each of the arcs BA, BC, is less than a semicircle,

and ∴ arc BA = arc BC.

Thus the arc ABC is bisected in B.

Q. E. F.

Ex. If, from any point in the diameter of a semicircle, there be drawn two straight lines to the circumference, one to the bisection of the circumference, and the other at right angles to the diameter, the squares on these two lines are together double of the square on the radius.

PROPOSITION XXXI. THEOREM.

In a circle, the angle in a semicircle is a right angle; and the angle in a segment greater than a semicircle is less than a right angle; and the angle in a segment less than a semicircle is greater than a right angle.

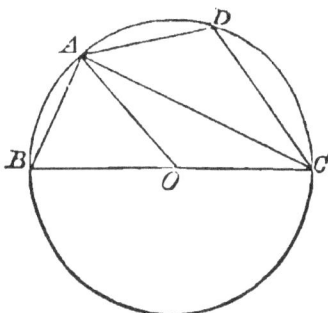

Let ABC be a ⊙, O its centre, and BC a diameter.

Draw AC, dividing the ⊙ into the segments ABC, ADC.
Join BA, AD, DC, AO.

Then must the ∠ in the semicircle BAC be a rt. ∠, and ∠ in segment ABC, greater than a semicircle, less than a rt. ∠, and ∠ in segment ADC, less than a semicircle, greater than a rt. ∠.

First, ∵ $BO = AO$, ∴ ∠ BAO = ∠ ABO;   I. A.

∴ ∠ COA = twice ∠ BAO;   I. 32.

and ∵ $CO = AO$, ∴ ∠ CAO = ∠ ACO;   I. A.

∴ ∠ BOA = twice ∠ CAO;   I. 32.

∴ sum of ∠s COA, BOA = twice sum of ∠s BAO, CAO, that is, two right angles = twice ∠ BAC.

∴ ∠ BAC is a right angle.

Next, ∵ ∠ BAC is a rt. ∠,

∴ ∠ ABC is less than a rt. ∠.   I. 17.

Lastly, ∵ sum of ∠s ABC, ADC = two rt. ∠s,   III. 22.

and ∠ ABC is less than a rt. ∠,

∴ ∠ ADC is greater than a rt. ∠.   Q. E. D.

Note.—For a simpler proof see page 178.

Ex. 1. If a circle be described on the radius of another circle as diameter, any straight line, drawn from the point, where they meet, to the outer circumference, is bisected by the interior one

Ex. 2. If a straight line be drawn to touch a circle, and be parallel to a chord, the point of contact will be the middle point of the arc cut off by the chord.

Ex. 3. If, from any point without a circle, lines be drawn touching it, the angle contained by the tangents is double of the angle contained by the line joining the points of contact, and the diameter drawn through one of them.

Ex. 4. The vertical angle of any oblique-angled triangle inscribed in a circle is greater or less than a right angle, by the angle contained by the base and the diameter drawn from the extremity of the base.

Ex. 5. If, from the extremities of any diameter of a given circle, perpendiculars be drawn to any chord of the circle that is not parallel to the diameter, the less perpendicular shall be equal to that segment of the greater, which is contained between the circumference and the chord.

Ex. 6. If two circles cut one another, and from either point of intersection diameters be drawn, the extremities of these diameters and the other point of intersection lie in the same straight line.

Ex. 7. Draw a straight line cutting two concentric circles, so that the part of it which is intercepted by the circumference of the greater may be twice the part intercepted by the circumference of the less.

PROPOSITION XXXII. THEOREM.

If a straight line touch a circle, and from the point of contact a straight line be drawn cutting the circle, the angles made by this line with the line touching the circle must be equal to the angles, which are in the alternate segments of the circle.

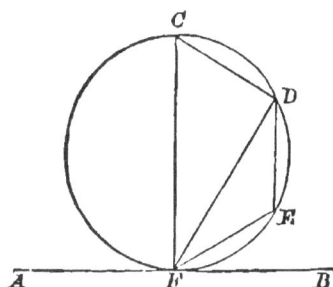

Let the st. line AB touch the $\odot$ $CDEF$ in F.
Draw the chord FD, dividing the $\odot$ into segments FCD, FED.
Then must $\angle DFB = \angle$ in segment FCD,
and $\angle DFA = \angle$ in segment FED.
From F draw the chord $FC \perp$ to AB.
Then FC is a diameter of the $\odot$.   III. 19.
Take any pt. E in the arc FED, and join FE, ED, DC.
Then $\because FDC$ is a semicircle, $\therefore \angle FDC$ is a rt. $\angle$;   III. 31.
$\therefore$ sum of $\angle$ s FCD, CFD = a rt. $\angle$.   I. 32.
Also, sum of $\angle$ s DFB, CFD = a rt. $\angle$.
$\therefore$ sum of $\angle$ s DFB, CFD = sum of $\angle$ s FCD, CFD,
and $\therefore \angle DFB = \angle FCD$,
that is, $\angle DFB = \angle$ in segment FCD.
Again, $\because CDEF$ is a quadrilateral fig. inscribed in a $\odot$,
$\therefore$ sum of $\angle$ s FED, FCD = two rt. $\angle$ s.   III. 22.
Also, sum of $\angle$ s DFA, DFB = two rt. $\angle$ s.   I. 13.
$\therefore$ sum of $\angle$ s DFA, DFB = sum of $\angle$ s FED, FCD ;
and $\angle DFB$ has been proved = $\angle FCD$;
$\therefore \angle DFA = \angle FED$,
that is, $\angle DFA = \angle$ in segment FED.

Q. E. D.

Ex. The chord joining the points of contact of parallel tangents is a diameter.

Proposition XXXIII. Problem.

On a given straight line to describe a segment of a circle capable of containing an angle equal to a given angle.

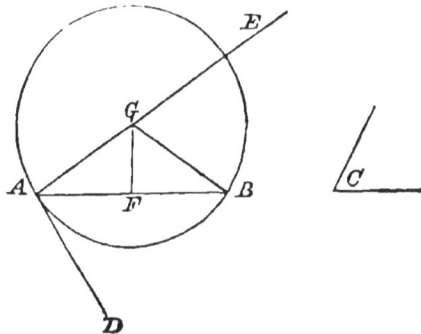

Let AB be the given st. line, and C the given $\angle$.

It is required to describe on AB a segment of a ⊙ which shall contain an $\angle = \angle C$.

At pt. A in st. line AB make $\angle BAD = \angle C$.　　I. 23.

Draw $AE \perp$ to AD, and bisect AB in F.

From F draw $FG \perp$ to AB, meeting AE in G.　Join GB.

Then in $\triangle$ s AGF, BGF ;

∵ $AF = BF$, and FG is common, and $\angle AFG = \angle BFG$;

∴ $GA = GB$.　　I. 4.

With G as centre and GA as radius describe a ⊙ ABH.

Then will AHB be the segment reqd.

For ∵ AD is $\perp$ to AE, a line passing through the centre,

∴ AD is a tangent to the ⊙ ABH.　　III. 16.

And ∵ the chord AB is drawn from the pt. of contact A,

∴ $\angle BAD = \angle$ in segment AHB,　　III. 32.

that is, the segment AHB contains an $\angle = \angle C$,

and it is described on AB, as was reqd.

Q. E. F.

Ex. 1. Two circles intersect in A, and through A is drawn a straight line meeting the circles again in P, Q. Prove that the angle between the tangents at P and Q is equal to the angle between the tangents at A.

Ex. 2. From two given points on the same side of a straight line, given in position, draw two straight lines which shall contain a given angle, and be terminated in the given line.

To cut off a segment from a given circle, capable of containing an angle equal to a given angle.

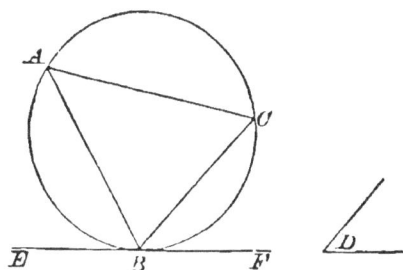

Let ABC be the given $\odot$, and D the given $\angle$.

It is required to cut off from $\odot$ ABC a segment capable of containing an $\angle = \angle D$.

Draw the st. line EBF to touch the circle at B.

At B make $\angle FBC = \angle D$.

Then $\because$ the chord BC is drawn from the pt. of contact B,

$\therefore \angle FBC = \angle$ in segment BAC,   III. 32.

that is, the segment BAC contains an $\angle = \angle D$;

and $\therefore$ a segment has been cut off from the $\odot$, as was reqd.

Q. E. F.

Ex. 1. If two circles touch internally at a point, any straight line passing through the point will divide the circles into segments, capable of containing equal angles.

Ex. 2. Given a side of a triangle, its vertical angle, and the radius of the circumscribing circle : construct the triangle.

Ex. 3. Given the base, vertical angle, and the perpendicular from the extremity of the base on the opposite side : construct the triangle.

Proposition XXXV. Theorem.

If two chords in a circle cut one another, the rectangle contained by the segments of one of them, is equal to the rectangle contained by the segments of the other.

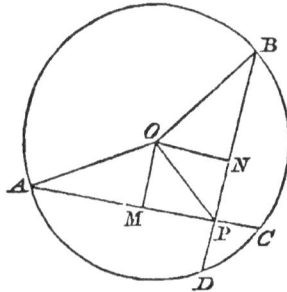

Let the chords AC, BD in the ⊙ $ABCD$ intersect in the pt. P.

Then must rect. AP, PC=rect. BP, PD.

From O, the centre, draw OM, ON ⊥ s to AC, BD,

and join OA, OB, OP.

Then ∵ AC is divided equally in M and unequally in P,

∴ rect. AP, PC with sq. on MP=sq. on AM.   II. 5.

Adding to each the sq. on MO,

rect. AP, PC with sqq. on MP, MO=sqq. on AM, MO ;

∴ rect. AP, PC with sq. on OP=sq. on OA.   I. 47.

In the same way it may be shewn that

rect. BP, PD with sq. on OP=sq. on OB.

Then ∵ sq. on OA=sq. on OB,

∴ rect. AP, PC with sq. on OP=rect. BP, PD with sq. on OP ;

∴ rect. AP, PC=rect. BP, PD.   Q. E. D.

Ex. 1. A and B are fixed points, and two circles are described passing through them ; PCQ, PCQ' are chords of these circles intersecting in C, a point in AB ; shew that the rectangle CP, CQ is equal to the rectangle CP', CQ'.

Ex. 2. If through any point in the common chord of two circles, which intersect one another, there be drawn any two other chords, one in each circle, their four extremities shall all lie in the circumference of a circle.

PROPOSITION XXXVI.   THEOREM.

If, from any point without a circle, two straight lines be drawn, one of which cuts the circle, and the other touches it ; the rectangle contained by the whole line which cuts the circle, and the part of it without the circle, must be equal to the square on the line which touches it.

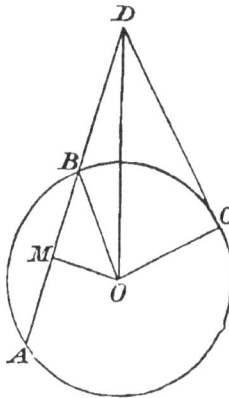

Let D be any pt. without the ⊙ ABC,
and let the st. lines DBA, DC be drawn to cut and touch the ⊙.
Then must rect. AD, DB=sq. on DC.
From O, the centre, draw OM bisecting AB in M,
and join OB, OC, OD.
Then ∵ AB is bisected in M and produced to D,
∴ rect. AD, DB with sq. on MB=sq. on MD.   II. 6.
Adding to each the sq. on MO,
rect. AD, DB with sqq. on MB, MO=sqq. on MD, MO.
Now the angles at M and C are rt. ∠ s ;   III. 3 and 18.
∴ rect. AD, DB with sq. on OB=sq. on OD ;
∴ rect. AD, DB with sq. on OB=sqq. on OC, DC.   I. 47.
And sq. on OB=sq. on OC ;
∴ rect. AD, DB=sq. on DC.   Q. E. D.

Proposition XXXVII. Theorem.

If, from a point without a circle, there be drawn two straight lines, one of which cuts the circle, and the other meets it ; if the rectangle contained by the whole line which cuts the circle, and the part of it without the circle, be equal to the square on the line which meets it, the line which meets must touch the circle.

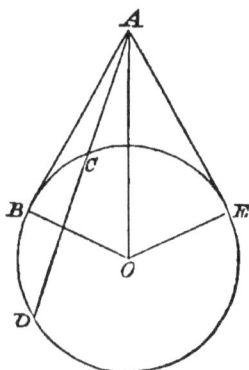

Let A be a pt. without the $\odot BCD$, of which O is the centre. From A let two st. lines ACD, AB be drawn, of which ACD cuts the $\odot$ and AB meets it.

Then if rect. DA, AC = sq. on AB, AB must touch the $\odot$.

Draw AE touching the $\odot$ in E, and join OB, OA, OE.

Then $\because ACD$ cuts the $\odot$, and AE touches it,

$\therefore$ rect. DA, AC = sq. on AE.   III. 36.

But rect. DA, AC = sq. on AB ;   Hyp.

$\therefore$ sq. on AB = sq. on AE ;

$\therefore AB = AE$.

Then in the $\triangle$s OAB, OAE,

$\because OB = OE$, and OA is common, and $AB = AE$,

$\therefore \angle ABO = \angle AEO$.   I. c.

But $\angle AEO$ is a rt. $\angle$;   III. 18.

$\therefore \angle ABO$ is a rt. $\angle$.

Now BO, if produced, is a diameter of the $\odot$;

$\therefore AB$ touches the $\odot$.   III. 16.

Q. E. D.

Miscellaneous Exercises on Book III.

1. The segments, into which a circle is cut by any straight line, contain angles, whose difference is equal to the inclination to each other of the straight lines touching the circle at the extremities of the straight line which divides the circle.

2. If from the point in which a number of circles touch each other, a straight line be drawn cutting all the circles, shew that the lines which join the points of intersection in each circle with its centre will be all parallel.

3. From a point Q in a circle, QN is drawn perpendicular to a chord PP', and QM perpendicular to the tangent at P : shew that the triangles NQP', QPM are equiangular.

4. AB, AC are chords of a circle, and D, E are the middle points of their arcs. If DE be joined, shew that it will cut off equal parts from AB, AC.

5. One angle of a quadrilateral figure inscribed in a circle is a right angle, and from the centre of the circle perpendiculars are drawn to the sides, shew that the sum of their squares is equal to twice the square of the radius.

6. A is the extremity of the diameter of a circle, O any point in the diameter. The chord which is bisected at O subtends a greater or less angle at A than any other chord through O, according as O and A are on the same or opposite sides of the centre.

7. If a straight line in a circle not passing through the centre be bisected by another and this by a third and so on, prove that the points of bisection continually approach the centre of the circle.

8. If a circle be described passing through the opposite angles of a parallelogram, and cutting the four sides, and the points of intersection be joined so as to form a hexagon, the straight lines thus drawn shall be parallel to each other.

9. If two circles touch each other externally and any third circle touch both, prove that the difference of the distances of

the centre of the third circle from the centres of the other two is invariable.

10. Draw two concentric circles, such that those chords of the outer circle, which touch the inner, may equal its diameter.

11. If the sides of a quadrilateral inscribed in a circle be bisected and the middle points of adjacent sides joined, the circles described about the triangles thus formed are all equal and all touch the original circle.

12. Draw a tangent to a circle which shall be parallel to a given finite straight line.

13. Describe a circle, which shall have a given radius, and its centre in a given straight line, and shall also touch another straight line, inclined at a given angle to the former.

14. Find a point in the diameter produced of a given circle, from which, if a tangent be drawn to the circle, it shall be equal to a given straight line.

15. Two equal circles intersect in the points A, B, and through B a straight line CBM is drawn cutting them again in C, M. Shew that if with centre C and radius BM a circle be described, it will cut the circle ABC in a point L such that arc AL=arc AB.

Shew also that LB is the tangent at B.

16. AB is any chord and AC a tangent to a circle at A ; CDE a line cutting the circle in D and E and parallel to AB. Shew that the triangle ACD is equiangular to the triangle EAB.

17. Two equal circles cut one another in the points A, B ; BC is a chord equal to AB ; shew that AC is a tangent to the other circle.

18. A, B are two points ; with centre B describe a circle, such that its tangent from A shall be equal to a given line.

19. The perpendiculars drawn from the angular points of a triangle to the opposite sides pass through the same point.

20. If perpendiculars be dropped from the angular points of a triangle on the opposite sides, shew that the sum of the squares on the sides of the triangle is equal to twice the sum of the rectangles, contained by the perpendiculars and that part of each intercepted between the angles of the triangles and the point of intersection of the perpendiculars.

21. When two circles intersect, their common chord bisects their common tangent.

22. Two circles intersect in A and B. Two points C and D are taken on one of the circles ; CA, CB meet the other circle in E, F, and DA, DB meet it in G, H : shew that FG is parallel to EH.

23. A and B are fixed points, and two circles are described passing through them ; CP, CP' are drawn from a point C on AB produced, to touch the circles in P, P' ; shew that $CP = CP'$.

24. From each angular point of a triangle a perpendicular is let fall upon the opposite side ; prove that the rectangles contained by the segments, into which each perpendicular is divided by the point of intersection of the three, are equal to each other.

25. If from a point without a circle two equal straight lines be drawn to the circumference and produced, shew that they will be at the same distance from the centre.

26. Let O, O' be the centres of two circles which cut each other in A, A'. Let B, B' be two points, taken one on each circumference. Let C, C' be the centres of the circles BAB', $BA'B'$. Then prove that the angle CBC' is the supplement of the angle $OA'O'$.

27. The common chord of two circles is produced to any point P ; PA touches one of the circles in A ; PBC is any chord of the other : shew that the circle which passes through A, B, C touches the circle to which PA is a tangent.

28. Given the base of a triangle, the vertical angle, and the length of the line drawn from the vertex to the middle point of the base : construct the triangle.

29. If a circle be described about the triangle ABC, and a straight line be drawn bisecting the angle BAC and cutting the circle in D, shew that the angle DCB will be equal to half the angle BAC.

30. If the line AD bisect the angle A in the triangle ABC, and BD be drawn without the triangle making an angle with BC equal to half the angle BAC, shew that a circle may be described about $ABCD$.

31. Two equal circles intersect in A, B : PQT perpendicular to AB meets it in T and the circles in P, Q. AP, BQ meet in R ; AQ, BP in S ; prove that the angle RTS is bisected by TF.

32. If the angle, contained by any side of a quadrilateral and the adjacent side produced, be equal to the opposite angle of the quadrilateral, prove that any side of the quadrilateral will subtend equal angles at the opposite angles of the quadrilateral.

33. If DE be drawn parallel to the base BC of a triangle ABC, prove that the circles described about the triangles ABC and ADE have a common tangent at A.

34. Describe a square equal to the difference of two given squares.

35. If tangents be drawn to a circle from any point without it, and a third line be drawn between the point and the centre of the circle, touching the circle, the perimeter of the triangle formed by the three tangents will be the same for all positions of the third point of contact. •

36. If on the sides of any triangle as chords, circles be described, of which the segments external to the triangle contain angles respectively equal to the angles of a given triangle, those circles will intersect in a point.

37. Prove that if ABC be a triangle inscribed in a circle, such that $BA = BC$, and AA' be drawn parallel to BC, meeting the circle again in A', and $A'B$ be joined cutting AC in E, BA touches the circle described about the triangle AEA'.

38. Describe a circle, cutting the sides of a given square, so that its circumference may be divided at the points of intersection into eight equal arcs.

39. *AB* is the diameter of a semicircle, *D* and *E* any two points on its circumference. Shew that if the chords joining *A* and *B* with *D* and *E*, either way, intersect in *F* and *G*, the tangents at *D* and *E* meet in the middle point of the line *FG*, and that *FG* produced is at right angles to *AB*.

40. Shew that the square on the tangent drawn from any point in the outer of two concentric circles to the inner equals the difference of the squares on the tangents, drawn from any point, without both circles, to the circles.

41. If from a point without a circle, two tangents *PT*, *PT'*, at right angles to one another, be drawn to touch the circle, and if from *T* any chord *TQ* be drawn, and from *T'* a perpendicular *T'M* be dropped on *TQ*, then *T'M = QM*.

42. Find the loci :

(1.) Of the centres of circles passing through two given points.

(2.) Of the middle points of a system of parallel chords in a circle.

(3.) Of points such that the difference of the distances of each from two given straight lines is equal to a given straight line.

(4.) Of the centres of circles touching a given line in a given point.

(5.) Of the middle points of chords in a circle that pass through a given point.

(6.) Of the centres of circles of given radius which touch a given circle.

(7.) Of the middle points of chords of equal length in a circle.

(8.) Of the middle points of the straight lines drawn from a given point to meet the circumference of a given circle.

43. If the base and vertical angle of a triangle be given, find the locus of the vertex.

44. A straight line remains parallel to itself while one of its extremities describes a circle. What is the locus of the other extremity ?

45. A ladder slips down between a vertical wall and a horizontal plane : what is the locus of its middle point ?

46. *ABC* is a line drawn from a point *A*, without a circle, to meet the circumference in *B* and *C*. Tangents are drawn to the circle at *B* and *C* which meet in *D*. What is the locus of *D* ?

47. The angular points *A, C* of a parallelogram *ABCD* move on two fixed straight lines *OA*, *OC*, whose inclination is equal to the angle *BCD* ; shew that one of the points *B, D*, which is the more remote from *O*, will move on a fixed straight line passing through *O*.

48. On the line *AB* is described the segment of a circle in the circumference of which any point *C* is taken. If *AC, BC* be joined, and a point *P* taken in *AC* so that *CP* is equal to *CB*, find the locus of *P*.

49. The centre of the circle *CBED* is on the circumference of *ABD*. If from any point *A* the lines *ABC* and *AED* be drawn to cut the circles, the chord *BE* is parallel to *CD*.

50. If a parallelogram be described having the diameter of a given circle for one of its sides, and the intersection of its diagonals on the circumference, shew that the extremity of each of the diagonals moves on the circumference of another circle of double the diameter of the first.

51. One diagonal of a quadrilateral inscribed in a circle is fixed, and the other of constant length. Shew that the sides will meet, if produced, on the circumferences of two fixed circles.

We here insert Euclid's proofs of Props. 23, 24 of Book III. first observing that he gives the following definition of similar segments :—

DEF. *Similar segments of circles are those in which the angles are equal, or which contain equal angles.*

PROPOSITION XXIII. THEOREM.

Upon the same straight line, and upon the same side of it, there cannot be two similar segments of circles, not coinciding with each other.

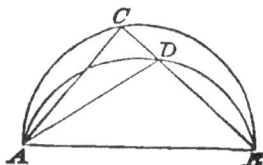

If it be possible, on the same base AB, and on the same side of it, let there be two similar segments of ⊙s, ABC, ABD, which do not coincide.

Because ⊙ ADB cuts ⊙ ACB in pts. A and B, they cannot cut one another in any other pt., and ∴ one of the segments must fall within the other.

Let ADB fall within ACB.

Draw the st. line BDC and join CA, DA.

Then ∵ segment ADB is similar to segment ACB,

∴ ∠ ADB = ∠ ACB.

Or the extr. ∠ of a △ = the intr. and opposite ∠, which is impossible ;

∴ the segments cannot but coincide.

Q. E. D.

Similar segments of circles, upon equal straight lines, are equal to one another.

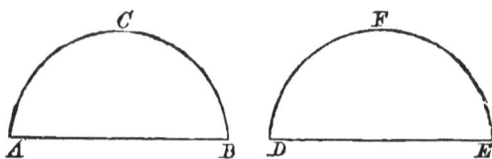

Let *ABC, DEF* be similar segments of ⊙s on equal st. lines *AB, DE*.

Then must segment *ABC*=segment *DEF*.

For if segment *ABC* be applied to segment *DEF*, so that *A* may be on *D* and *AB* on *DE*, then *B* will coincide with *E*, and *AB* with *DE* ;

∴ segment *ABC* must also coincide with segment *DEF* ;

III. 23.

∴ segment *ABC*=segment *DEF*.　　Ax. 8.

Q. E. D.

We gave one Proposition, C, page 150, as an example of the way in which the conceptions of Flat and Reflex Angles may be employed to extend and simplify Euclid's proofs. We here give the proofs, based on the same conceptions, of the important propositions XXII. and XXXI.

The opposite angles of any quadrilateral figure, inscribed in a circle, are together equal to two right angles.

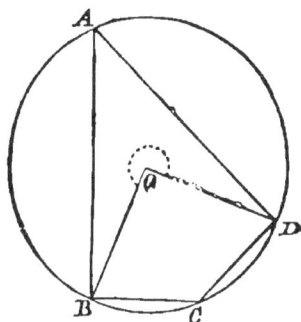

Let $ABCD$ be a quadrilateral fig. inscribed in a ⊙.

Then must each pair of its opposite ∠ s be together equal to two rt. ∠ s.

From O, the centre, draw OB, OD.

Then ∵ ∠ BOD = twice ∠ BAD, III. 20.

and the reflex ∠ DOB = twice ∠ BCD, III. C. p. 150.

∴ sum of ∠ s at O = twice sum of ∠ s BAD, BCD.

But sum of ∠ s at O = 4 right ∠ s ; I. 15, Cor. 2.

∴ twice sum of ∠ s BAD, BCD = 4 right ∠ s ;

∴ sum of ∠ s BAD, BCD = two right ∠ s.

Similarly, it may be shewn that

sum of ∠ s ABC, ADC = two right ∠ s.

Q. E. D.

PROPOSITION XXXI. THEOREM.

In a circle, the angle in a semicircle is a right angle; and the angle in a segment greater than a semicircle is less than a right angle; and the angle in a segment less than a semicircle is greater than a right angle.

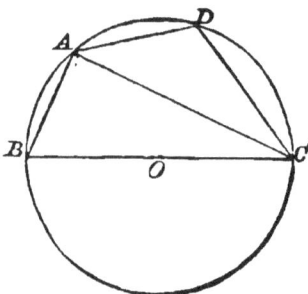

Let ABC be a $\odot$, of which O is the centre and BC a diameter.

Draw AC, dividing the $\odot$ into the segments ABC, ADC.

Join BA, AD, DC.

Then must the $\angle$ in the semicircle BAC be a rt. $\angle$, and $\angle$ in segment ABC, greater than a semicircle, less than a rt. $\angle$, and $\angle$ in segment ADC, less than a semicircle, greater than a rt. $\angle$.

First, $\because$ the flat angle $BOC =$ twice $\angle BAC$,   III. C. p. 150.

$\therefore \angle BAC$ is a rt. $\angle$.

Next, $\because \angle BAC$ is a rt. $\angle$,

$\therefore \angle ABC$ is less than a rt. $\angle$.   I. 17.

Lastly, $\because$ sum of $\angle$ s ABC, $ADC =$ two rt. $\angle$ s,   III. 22.

and $\angle ABC$ is less than a rt. $\angle$,

$\therefore \angle ADC$ is greater than a rt. $\angle$.

Q. E. D.

BOOK IV.

INTRODUCTORY REMARKS.

EUCLID gives in this Book of the Elements a series of Problems relating to cases in which circles may be described in or about triangles, squares, and regular polygons, and of the last-mentioned he treats of three only :

the Pentagon, or figure of 5 sides,

„ Hexagon, „ 6 „

„ Quindecagon, „ 15 „ .

The Student will find it useful to remember the following Theorems, which are established and applied in the proofs of the Propositions in this Book.

I. The bisectors of the angles of a triangle, square, or regular polygon meet in a point, which is the centre of the inscribed circle.

II. The perpendiculars drawn from the middle points of the sides of a triangle, square, or regular polygon meet in a point, which is the centre of the circumscribed circle.

III. In the case of a square, or regular polygon the inscribed and circumscribed circles have a common centre.

IV. If the circumference of a circle be divided into any number of equal parts, the chords joining each pair of consecutive points form a regular figure inscribed in the circle, and the tangents drawn through the points form a regular figure described about the circle.

Proposition I. Problem.

In a given circle to draw a chord equal to a given straight line, which is not greater than the diameter of the circle.

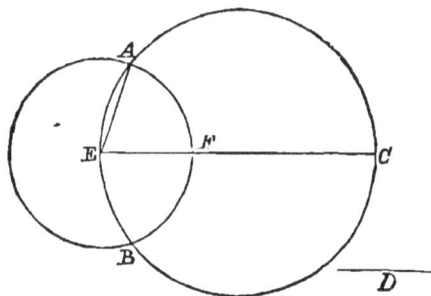

Let ABC be the given $\odot$, and D the given line, not greater than the diameter of the $\odot$.

It is required to draw in the $\odot$ ABC a chord $=D$.

Draw EC, a diameter of $\odot$ ABC.

Then if $EC=D$, what was required is done.

But if not, EC is greater than D. From EC cut off $EF=D$, and with centre E and radius EF describe a $\odot$ AFB, cutting the $\odot$ ABC in A and B; and join AE.

Then, $\because$ E is the centre of $\odot$ AFB,

$$\therefore EA=EF,$$

and $\therefore EA=D$.

Thus a chord EA equal to D has been drawn in $\odot$ ABC.

Q. E. F.

Ex. Draw the diameter of a circle, which shall pass at a given distance from a given point.

In a given circle to inscribe a triangle, equiangular to a given triangle.

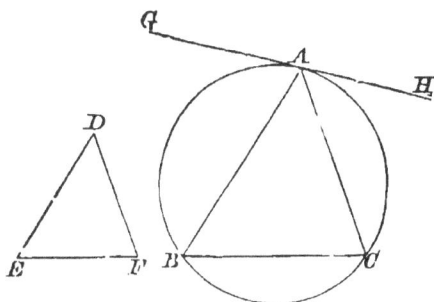

Let ABC be the given $\odot$, and DEF the given $\triangle$.

It is required to inscribe in $\odot$ ABC a $\triangle$, equiangular to $\triangle$ DEF.

Draw GAH touching the $\odot$ ABC at the pt. A.   III. 17.

Make $\angle GAB = \angle DFE$, and $\angle HAC = \angle DEF$.   I. 23.

Join BC. Then will $\triangle$ ABC be the required $\triangle$.

For $\because$ GAH is a tangent, and AB a chord of the $\odot$,

$$\therefore \angle ACB = \angle GAB,$$ III. 32.

that is, $\angle ACB = \angle DFE$.

So also, $\angle ABC = \angle HAC$,   III. 32.

that is, $\angle ABC = \angle DEF$;

$\therefore$ remaining $\angle BAC =$ remaining $\angle EDF$;

$\therefore$ $\triangle$ ABC is equiangular to $\triangle$ DEF, and it is inscribed in the $\odot$ ABC.

Q. E. F.

Ex. If an equilateral triangle be inscribed in a circle, prove that the radii, drawn to the angular points, bisect the angles of the triangle.

PROPOSITION III.   PROBLEM.

About a given circle to describe a triangle, equiangular to a given triangle.

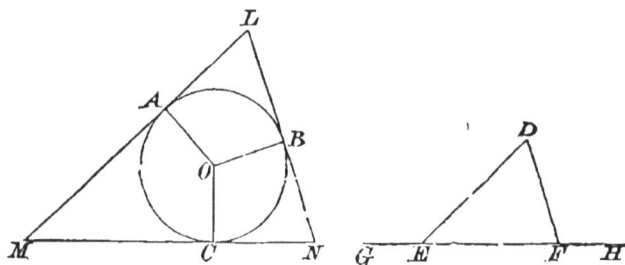

Let ABC be the given $\odot$, and DEF the given $\triangle$.

It is required to describe about the $\odot$ a $\triangle$ equiangular to $\triangle$ EDF.

From O, the centre of the $\odot$, draw any radius OC.
Produce EF to the pts. G, H.
Make $\angle COA = \angle DEG$, and $\angle COB = \angle DFH$.   I. 23.
Through A, B, C draw tangents to the $\odot$, meeting in L, M, N.
Then will LMN be the $\triangle$ required.
For $\because$ ML, LN, NM are tangents to the $\odot$,
$\therefore$ the $\angle$ s at A, B, C are rt. $\angle$ s.   III. 18.
Now $\angle$ s of quadrilateral $AOCM$ together = four rt. $\angle$ s. ;
and of these $\angle OAM$ and $\angle OCM$ are rt. $\angle$ s ;
$\therefore$ sum of $\angle$ s COA, AMC = two rt. $\angle$ s.
But sum of $\angle$ s DEG, DEF = two rt. $\angle$ s ;   I. 32.
$\therefore$ sum of $\angle$ s COA, AMC = sum of $\angle$ s DEG, DEF.
and $\angle COA = \angle DEG$, by construction ;
$\therefore$ $\angle AMC = \angle DEF$;
that is $\angle LMN = \angle DEF$.
Similarly, it may be shewn that $\angle LNM = \angle DFE$;
$\therefore$ also $\angle MLN = \angle EDF$.
Thus a $\triangle$, equiangular to $\triangle$ DEF, is described about the $\odot$.

Q. E. F.

PROPOSITION IV. PROBLEM.

To inscribe a circle in a given triangle.

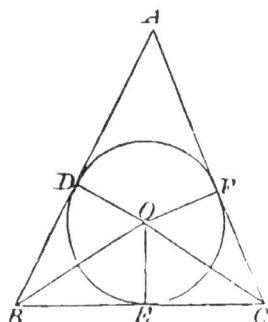

Let *ABC* be the given △.

It is required to inscribe a ⊙ in the △ ABC.

Bisect ∠ s *ABC*, *ACB* by the st. lines *BO*, *CO*, meeting in *O*.   I. 9.

From *O* draw *OD*, *OE*, *OF*, ⊥ s to *AB*, *BC*, *CA*.   I. 12.

Then, in △ s *EBO*, *DBO*,

∵ ∠ *EBO* = ∠ *DBO*, and ∠ *BEO* = ∠ *BDO*, and *OB* is common,

∴ *OE* = *OD*.   I. 26.

Similarly it may be shewn that *OE* = *OF*.

If then a ⊙ be described, with centre *O*, and radius *OD*, this ⊙ will pass through the pts. *D*, *E*, *F* ;

and ∵ the ∠ s at *D*, *E* and *F* are rt. ∠ s,

∴ *AB*, *BC*, *CA* are tangents to the ⊙ ;   III. 16.

and thus a ⊙ *DEF* may be inscribed in the △ *ABC*.

Q. E. F.

Ex. 1. Shew that, if *OA* be drawn, it will bisect the angle *BAC*.

Ex. 2. If a circle be inscribed in a right-angled triangle, the difference between the hypotenuse and the sum of the other sides is equal to the diameter of the circle.

Ex. 3. Shew that, in an equilateral triangle, the centre of the inscribed circle is equidistant from the three angular points.

Ex. 4. Describe a circle, touching one side of a triangle and the other two produced. (NOTE. This is called an *escribed* circle.)

NOTE. Euclid's fifth Proposition of this Book has been already given on page 135.

PROPOSITION VI. PROBLEM.

To inscribe a square in a given circle.

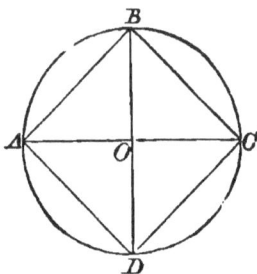

Let $ABCD$ be the given ⊙.

It is required to inscribe a square in the ⊙.

Through O, the centre, draw the diameters AC, BD, ⊥ to each other.

Join AB, BC, CD, DA.

Then ∵ the ∠ s at O are all equal, being rt. ∠ s,   I. Post. 4.

· the area AD, BC, CD, DA are all equal,   III. 26.

and ∴ the chords AB, BC, CD, DA are all equal ;   III. 29.

and ∠ ABC, being the ∠ in a semicircle, is a rt. ∠ . III. 31.

So also the ∠ s BCD, CDA, DAB are rt. ∠ s ;

∴ $ABCD$ is a square,

and it is inscribed in the ⊙ as was required.

Q. E. F.

To describe a square about a given circle.

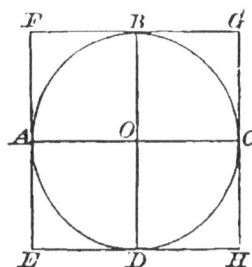

Let $ABCD$ be the given $\odot$, of which O is the centre.

It is required to describe a square about the $\odot$.

Draw the diameters AC, BD, $\perp$ to each other.
Through A, B, C, D draw EF, FG, GH, HE
touching the $\odot$.   III. 17.

     Then the $\angle$ s at A, B, C, D are rt. $\angle$ s.   III. 16.

     Now $\because$ the $\angle$ s at A, O, C are all rt. $\angle$ s,

         $\therefore$ FE, BD, and GH are all $\parallel$;   I. 27.

     and $\because$ the $\angle$ s at B, O, D are all rt. $\angle$ s,

         $\therefore$ FG, AC, and EH are all $\parallel$;

         $\therefore$ FE and GH each $= BD$,   I. 34.

     and FG and EH each $= AC$.   I. 34.

       And $\because$ $BD = AC$,

$\therefore$ FE, GH, FG, EH, are all equal.

     Again, $\because$ FO is a $\square$,

         $\therefore \angle AFB = \angle AOB$,   I. 34.

     and $\therefore \angle AFB$ is a rt. $\angle$.

     So also the $\angle$ s at G, H, and E are rt. $\angle$ s.

Hence $EFGH$ is a square, and it is described about the $\odot$.

                             Q. E. F.

EX. In a given circle inscribe four circles, equal to each
other, and in mutual contact with each other and with the
given circle.

PROPOSITION VIII. PROBLEM.

To inscribe a circle in a given square.

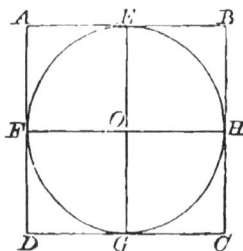

Let *ABCD* be the given square.

It is required to inscribe a ⊙ in the square.

Bisect *AB*, *AD* in *E*, *F*,   I. 10.

and draw *EG* ‖ to *AD* or *BC*, and *FH* ‖ to *AB* or *DC*.

Let *EG* and *FH* intersect in *O*.

Then ∵ *AO* is a ▱,

∴ *OE=FA* and *OF=EA*.   I. 34.

But ∵ *AB=AD*, and *E*, *F* are the middle pts. of *AB*, *AD*,

∴ *FA=EA*,

and ∴ *OE=OF*.

Similarly, it may be shewn that *OG=OF*, and *OH=OE*,

and ∴ *OE*, *OF*, *OG*, *OH* are all equal ;

and a ⊙, described with centre *O* and radius *OE*,

will pass through *E*, *F*, *G*, *H*,

and it will be touched by each of the sides of the square,

. the ∠ s at *E*, *F*, *G*, *H* are rt. ∠ s.   III. 16.

Thus a ⊙ *EFGH* may be inscribed in the sq. *ABCD*.

Q. E. F.

Ex. 1. In what parallelograms can circles be inscribed ?

Ex. 2. If, from any point in the circumference of a circle, straight lines be drawn to the angular points of the inscribed square, the sum of the squares on these four lines will be double of the square on the diameter.

To describe a circle about a given square.

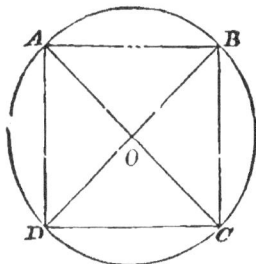

Let $ABCD$ be the given square.

It is required to describe a ⊙ about the square.

Draw the diagonals AC, BD, intersecting each other in O.

Then ∵ $\angle DAC = \angle ACD$,   I. A.

and $\angle BAC =$ alternate $\angle ACD$,   I. 29.

∴ $\angle DAC = \angle BAC$.

Thus the diagonal AC bisects $\angle BAD$,

and ∴ $\angle OAB =$ half a rt. $\angle$.

Similarly it may be shewn that $\angle OBA =$ half a rt. $\angle$;

∴ $\angle OBA = \angle OAB$;

∴ $OA = OB$.   I. B. Cor.

Similarly it may be shewn that $OC = OB$, and $OD = OA$;

∴ OA, OB, OC, OD are all equal;

and ∴ a ⊙, described with centre O and radius OA, will pass through A, B, C, D, and will be described about the square, as was required.

Q. E. F.

PROPOSITION X.   PROBLEM.

To describe an isosceles triangle, having each of the angles at the base double of the third angle.

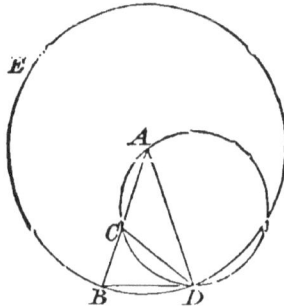

Take any st. line AB and divide it in C,
      so that rect. AB, $BC =$ sq. on AC.   II. 11.
With centre A and radius AB describe the $\odot$ BDE,
and in it draw the chord $BD = AC$; and join AD.   IV. 1.
    Then will $\triangle$ ABD *have each of the* $\angle$ *s at the base double of* $\angle$ BAD.
    Join CD, and about the $\triangle ACD$ describe the $\odot ACD$.   IV. 5.
    Then $\because$ rect. AB, $BC =$ sq. on AC, and $BD = AC$,
      $\therefore$ rect. AB, $BC =$ sq. on BD,
      and $\therefore$ BD touches the $\odot$ ACD.   III. 37.
    Then $\because$ BD touches $\odot$ ACD, and DC is a chord of the $\odot$
          $\therefore \angle BDC = \angle CAD$.   III. 32.
          Add to each $\angle CDA$.
      Then $\angle BDA =$ sum of $\angle$ s CAD, CDA,
          $\therefore \angle BDA = \angle BCD$.   I. 32.
      But $\angle BDA = \angle CBD$;   I. A.
          $\therefore \angle BCD = \angle CBD$,
          and $\therefore$ $BD = UD$.   I. B. Cor.
      But $BD = CA$;
          $\therefore CA = CD$,
      and $\therefore \angle CDA = \angle CAD$.   I. A.
    Hence sum of $\angle$ s CDA, $CAD =$ twice $\angle CAD$,
          $\therefore \angle BCD =$ twice $\angle BAD$.   I. 32.
    But $\angle ABD$ and $\angle ADB$ are each $= \angle BCD$,
      $\therefore \angle ABD$ and $\angle ADB$ are each $=$ twice $\angle BAD$;
    and thus an isosceles $\triangle$ ABD has been described as was required.   Q. E. F.

To inscribe a regular pentagon in a given circle.

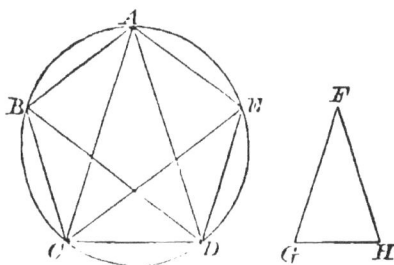

Let $ABCDE$ be the given ⊙.

It is required to inscribe a regular pentagon in the ⊙.

Make an isosceles △ FGH, having each of the ∠s at G, H double of ∠ at F.

In ⊙ $ABCDE$ inscribe a △ ACD equiangular to △ FGH,   IV. 2.

having ∠s at A, C, D = the ∠s at F, G, H, respectively.

Then ∠ ADC = twice ∠ DAC, and ∠ ACD = twice ∠ DAC.

Bisect the ∠s ADC, ACD by the chords DB, CE.

Join AB, BC, DE, EA.

Then will $ABCDE$ be a regular pentagon.

For ∵ ∠s ADC, ACD are each = twice ∠ DAC,

and ∠s ADC, ACD are bisected by DB, CE,

∴ ∠s ADB, BDC, DAC, ECD, ACE, are all equal ;

and ∴ arcs AB, BC, CD, DE, EA are all equal ;   III. 26.

and ∴ chords AB, BC, CD, DE, EA are all equal.   III. 29.

Hence, the pentagon $ABCDE$ is equilateral.

Again, ∵ arc CD = arc AB,

adding to each arc AED, we have

arc $AEDC$ = arc $BAED$,

and ∴ ∠ ABC = ∠ BCD.   III. 27.

Similarly, ∠s CDE, DEA, EAB each = ∠ ABC.

Hence, the pentagon $ABCDE$ is equiangular.

Thus a regular pentagon has been inscribed in the ⊙.

Q. E. F.

Ex. Shew that CE is parallel to BA.

To describe a regular pentagon about a given circle.

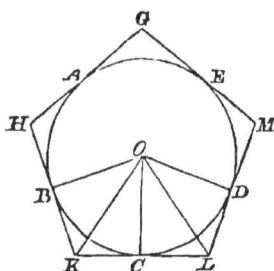

Let $ABCDE$ be the given ⊙.

It is required to describe a regular pentagon about the ⊙.

Let the angular pts. of a regular pentagon inscribed in the ⊙ be at A, B, C, D, E,

so that the arcs AB, BC, CD, DE, EA are all equal.

Through A, B, C, D, E draw GH, HK, KL, LM, MG tangents to the ⊙;

take the centre O, and join OB, OK, OC, OL, OD.

Then in △s OBK, OCK,

∵ $OB = OC$, and OK is common, and $KB = KC$,

I. E. Cor.

∴ ∠ BKO = ∠ CKO, and ∠ BOK = ∠ COK,

that is, ∠ BKC = twice ∠ CKO, and ∠ BOC = twice ∠ COK.

So also, ∠ DLC = twice ∠ CLO, and ∠ DOC = twice ∠ COL.

Now ∵ arc BC=arc CD,

$$\therefore \angle BOC = \angle DOC,$$

and $\therefore \angle COK = \angle COL.$

Hence in △ s OCK, OCL,

∵ $\angle COK = \angle COL$, and rt. $\angle OCK$=rt. $\angle OCL$, and OC is common,

$$\therefore \angle CKO = \angle CLO, \text{ and } CK = CL, \qquad \text{I. b.}$$

and $\therefore \angle HKL = \angle MLK$, and KL=twice KC.

Similarly it may be shewn that $\angle$ s KHG, HGM, GML each = $\angle HKL$,

$\therefore$ the pentagon $GHKLM$ is equiangular.

And since it has been shewn that KL=twice KC,

and it can be shewn that HK=twice KB,

and ∵ $KB = KC$,   I. E. Cor.

$$\therefore HK = KL.$$

In like manner it may be shewn that HG, GM, ML, each = KL,

$\therefore$ the pentagon $GHKLM$ is equilateral.

Thus a regular pentagon has been described about the ⊙.

Q. E. F.

PROPOSITION XIII.   PROBLEM.

To inscribe a circle in a given regular pentagon.

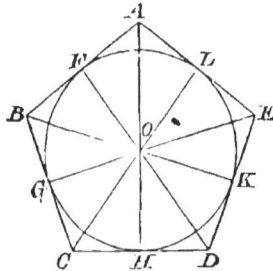

Let $ABCDE$ be the given regular pentagon.

It is required to inscribe a ⊙ in the pentagon.

Bisect ∠ s BCD, CDE by the st. lines CO, DO, meeting in O.

Join OB, OA, OE.

Then, in △ s BCO, DCO,

∵ $BC = DC$, and CO is common, and ∠ $BCO = $ ∠ DCO,

∴ ∠ $OBC = $ ∠ ODC.   I. 4.

Then, ∵ ∠ $ABC = $ ∠ CDE,   Hyp.

and ∠ $CDE =$ twice ∠ ODC,

∴ ∠ $ABC =$ twice ∠ OBC.

Hence OB bisects ∠ ABC.

In the same way we can shew that OA, OE bisect
the ∠ s BAE, AED.

Draw OF, OG, OH, OK, OL ⊥ to AB, BC, CD, DE, EA.

Then, in △ s GOC, HOC,

∵ ∠ $GCO = $ ∠ HCO, and ∠ $OGC = $ ∠ OHC,

and OC is common,

∴ $OG = OH$.   I. 26.

So also it may be shewn that OF, OL, OK are
each $= OG$ or OH ;

∴ OF, OG, OH, OK, OL are all equal.

Hence a ⊙ described with centre O and radius OF
will pass through G, H, K, L,
and will touch the sides of the pentagon,
∵ the ∠ s at F, G, H, K, L are rt. ∠ s.   III. 16.

Thus a ⊙ will be inscribed in the pentagon.   Q. E. F.

PROPOSITION XIV.   PROBLEM.

To describe a circle about a given regular pentagon.

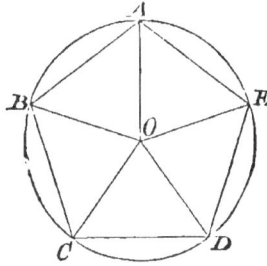

Let *ABCDE* be the given regular pentagon.

It is required to describe a ⊙ about the pentagon.

Bisect the ∠s *BCD*, *CDE* by the st. lines *CO*, *DO*, meeting in *O*.

Join *OB*, *OA*, *OE*.

Then it may be shewn, as in the preceding Proposition, that

OB, *OA*, *OE* bisect the ∠s *CBA*, *BAE*, *AED*.

And ∵ ∠ *BCD* = ∠ *CDE*,

and ∠ *OCD* = half ∠ *BCD*, and ∠ *ODC* = half ∠ *CDE*,

∴ ∠ *OCD* = ∠ *ODC*,

and ∴ *OD* = *OC*.

In the same way we may shew that *OB*, *OA*, *OE*

each = *OD* or *OC* ;

∴ *OA*, *OB*, *OC*, *OD*, *OE* are all equal,

and a ⊙ described with centre *O* and radius *OA* will pass through *B*, *C*, *D*, *E*,

and will be described about the pentagon.

Q. E. F.

PROPOSITION XV. PROBLEM.

To inscribe a regular hexagon in a given circle.

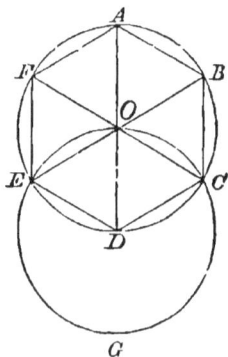

Let $ABCDEF$ be the given $\odot$, of which O is the centre.

It is required to inscribe a regular hexagon in the $\odot$.

Draw the diameter AOD,

and with centre D and radius DO describe a $\odot$ $EOCG$

Join EO, CO, and produce them to B and F.

Join AB, BC, CD, DE, EF, FA.

Then $\because$ O is the centre of $\odot$ ACE, $\therefore$ $OE = OD$;

and $\because$ D is the centre of $\odot$ GCE, $\therefore$ $OD = DE$;

$\therefore$ OED is an equilateral $\triangle$,

and $\therefore$ $\angle EOD$ = the third part of two rt. $\angle$ s.   I. 32

So also $\angle DOC$ = the third part of two rt. $\angle$ s,

and $\therefore$ $\angle BOC$ = the third part of two rt. $\angle$ s.   I. 13

Thus $\angle$ s EOD, DOC, BOC are all equal;

and to these the vertically opposite $\angle$ s BOA, AOF, FOE

are equal;   I. 15.

$\therefore$ $\angle$ s AOB, BOC, COD, DOE, EOF, FOA, are all equal,

and $\therefore$ arcs AB, BC, CD, DE, EF, FA are all equal.

    III. 26.

and $\therefore$ chords AB, BC, CD, DE, EF, FA are all equal.

    III. 29.

Thus the hexagon $ABCDEF$ is equilateral.

Also $\because$ each of its $\angle$ s = two-thirds of two rt. $\angle$ s,

$\therefore$ the hexagon $ABCDEF$ is equiangular.

Thus a regular hexagon has been inscribed in the $\odot$.

    Q. E. F.

PROPOSITION XVI.   PROBLEM.

To inscribe a regular quindecagon in a given circle.

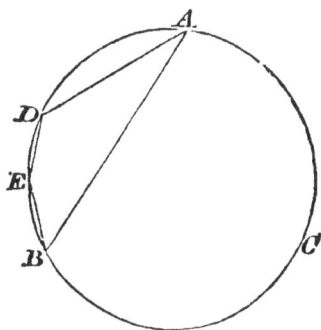

Let *ABC* be the given ⊙.

It is required to inscribe in the ⊙ a regular quindecagon.

Let *A B* be the side of an equilateral △ inscribed in the ⊙,

IV. 2.

and *AD* the side of a regular pentagon inscribed in the ⊙.

IV. 11.

Then of such equal parts as the whole ○ce *ABC* contains fifteen,

arc *ADB* must contain five,

and arc *AD* must contain three,

and ∴ arc *DB*, their difference, must contain two. ·

Bisect arc *DB* in *E*.   III. 30.

Then arcs *DE*, *EB* are each the fifteenth part of the whole ○ce.

If then chords *DE*, *EB* be drawn,

and chords equal to them be placed all round the ○ce,   IV. 1.

a regular quindecagon will be inscribed in the ⊙.

Q. E. F.

Miscellaneous Exercises on Book IV.

1. The perpendiculars let fall on the sides of an equilateral triangle from the centre of the circle, described about tha triangle, are equal.

2. Inscribe a circle in a given regular octagon.

3. Shew that in the diagram of Prop. X. there is a second triangle, which has each of two of its angles double of the third.

4. Describe a circle about a given rectangle.

5. Shew that the diameter of the circle which is described about an isosceles triangle, which has its vertical angle double of either of the angles at the base, is equal to the base of the triangle.

6. The side of the equilateral triangle, described about a circle, is double of the side of the equilateral triangle, inscribed in the circle.

7. A quadrilateral figure may have a circle described about it, if the rectangles contained by the segments of the diagonals be equal.

8. The square on the side of an equilateral triangle, inscribed in a circle, is triple of the square on the side of the regular hexagon, inscribed in the same circle.

9. Inscribe a circle in a given rhombus.

10. *ABC* is an equilateral triangle inscribed in a circle ; tangents to the circle at *A* and *B* meet in *M*. Shew that a diameter drawn from *M*, and meeting the circumference in *D* and *C*, bisects the angle *AMB*, and that *DC* is equal to twice *MD*.

11. Compare the areas of two regular hexagons, one inscribed in, the other described about, a given circle.

12. Inscribe a square in a given semicircle.

13. A circle being given, describe six other circles, each of them equal to it, and in contact with each other and with the given circle.

14. Given the angles of a triangle, and the perpendiculars from any point on the three sides, construct the triangle.

15. Having given the radius of a circle, determine its centre, when the circle touches two given lines, which are not parallel.

16. If the distance between the centres of two circles, which cut one another at right angles, is equal to twice one of the radii, the common chord is the side of the regular hexagon, inscribed in one of the circles, and the side of the equilateral triangle, inscribed in the other.

17. If from O, the centre of the circle inscribed in a triangle ABC, OD, OE, OF be drawn perpendicular to the sides BC, CA, AB, respectively, and from any point P in OP, drawn parallel to AB, perpendiculars PQ, PR be drawn upon OD and OE respectively, or these produced, shew that the triangle QRO is equiangular to the triangle ABC.

Euclid Papers set in the Mathematical Tripos at Cambridge from 1848 to 1872.

QUESTIONS arising out of the Propositions, to which they are attached, have been proposed in the Euclid Papers to Candidates for Mathematical Honours since the year 1848.

A complete set of these questions, so far as they refer to Books I.-IV., is here given. The figures preceding each question denote the particular Proposition to which the question was attached. It is expected that the solution of each question is to be obtained mainly by using the Proposition which precedes it, and that no Proposition which comes later in Euclid's order should be assumed.

Of some of the questions here given we have already made use in the preceding pages. As examples, however, of what has been hitherto expected of Candidates for Honours, and in order to keep the series of Papers complete, we have not hesitated to repeat them.

1848.   I. 34. If the two diagonals be drawn, shew that a parallelogram will be divided into four equal parts. In what case will the diagonal bisect the angles of the parallelogram?

III. 16. Shew that all equal straight lines in a circle will be touched by another circle.

III. 20. If two straight lines *AEB, CED* in a circle intersect in *E*, the angles subtended by *AC* and *BD* at the centre are together double of the angle *AEC*.

1849.   I. 1. By a method similar to that used in this problem, describe on a given finite straight line an isosceles triangle, the sides of which shall be each equal to twice the base.

      II. 11. Shew that in Euclid's figure four other lines beside the given line, are divided in the required manner.

      IV. 4. Describe a circle touching one side of a triangle and the produced parts of the other two.

1850.   I. 34. If the opposite sides, or the opposite angles, of any quadrilateral figure be equal, or if its diagonals bisect each other, the quadrilateral is a parallelogram.

      II. 14. Given a square, and one side of a rectangle which is equal to the square, find the other side.

      III. 31. The greatest rectangle that can be inscribed in a circle is a square.

      III. 34. Divide a circle into two segments such that the angle in one of them shall be five times the angle in the other.

      IV. 10. Shew that the base of the triangle is equal to the side of a regular pentagon inscribed in the smaller circle of the figure.

1851.   I. 38. Let ABC, ABD be two equal triangles, upon the same base AB and on opposite sides of it: join CD, meeting AB in E: shew that CE is equal to ED.

      I. 47. If ABC be a triangle, whose angle A is a right angle, and BE, CF be drawn bisecting the opposite sides respectively, shew that four times the sum of the squares on BE and CF is equal to five times the square on BC.

      III. 22. If a polygon of an even number of sides be inscribed in a circle, the sum of the alternate angles together with two right angles is equal to as many right angles as the figure has sides.

1851.   IV. 16. In a given circle inscribe a triangle, whose
angles are as the numbers 2, 5 and 8.

1852.   I. 42. Divide a triangle by two straight lines into
three parts, which, when properly arranged,
shall form a parallelogram whose angles are
of given magnitude.

II. 12. Triangles are described on the same base and
having the difference of the squares on the
other sides constant : shew that the vertex of
any triangle is in one or other of two fixed
straight lines.

IV. 3. Two equilateral triangles are described about
the same circle : shew that their intersections
will form a hexagon equilateral, but not generally equiangular.

1853. I. B. Cor. If lines be drawn through the extremities of the
base of an isosceles triangle, making angles
with it, on the side remote from the vertex,
each equal to one third of one of the equal
angles, and meeting the sides produced, prove
that three of the triangles thus formed are
isosceles.

I. 29. Through two given points draw two lines, forming with a line, given in position, an equilateral triangle.

II. 11. In the figure, if H be the point of division of
the given line AB, and DA be the side of the
square which is bisected in E and produced
to F, and if DH be produced to meet BF in
L, prove that DL is perpendicular to BF, and
is divided by BE similarly to the given line.

III. 32. Through a given point without a circle draw a
chord such that the difference of the angles
in the two segments, into which it divides the
circle, may be equal to a given angle.

III. 36. From a given point as centre describe a circle cutting a given line in two points, so that the rectangle contained by their distances from a fixed
point in the line may be equal to a given square

1854. ı. 43. If K be the common angular point of the parallelograms about the diameter, and BD the other diameter, the difference of the parallelograms is equal to twice the triangle BKD.

ıı. 1ı. Produce a given straight line to a point such that the rectangle contained by the whole line thus produced and the part produced shall be equal to the square on the given straight line.

ııı. 22. If the opposite sides of the quadrilateral be produced to meet in P, Q, and about the triangles so formed without the quadrilateral circles be described meeting again in R, shew that P, R, Q will be in one straight line.

ıv. 10. Upon a given straight line, as base, describe an isosceles triangle having the third angle treble of each of the angles at the base.

1855. ı. 20. Prove that the sum of the distances of any point from the three angles of a triangle is greater than half the perimeter of the triangle.

ı. 47. If a line be drawn parallel to the hypotenuse of a right-angled triangle, and each of the acute angles be joined with the points where this line intersects the sides respectively opposite to them, the squares on the joining lines are together equal to the squares on the hypotenuse and on the line drawn parallel to it.

ıı. 9. Divide a given straight line into two parts, such that the square on one of them may be double of the square on the other, without employing the Sixth Book.

ııı. 27. If any number of triangles, upon the same base BC, and on the same side of it, have their vertical angles equal, and perpendiculars meeting in D be drawn from B, C upon the opposite sides, find the locus of D, and shew that all the lines which bisect the angle BDC pass through the same point.

1855. IV. 4. If the circle inscribed in a triangle ABC touch the sides AB, AC in the points D, E, and a straight line be drawn from A to the centre of the circle, meeting the circumference in G, shew that G is the centre of the circle inscribed in the triangle ADE.

1856. I. 34. Of all parallelograms, which can be formed with diameters of given length, the rhombus is the greatest.

II. 12. If AB, one of the equal sides of an isosceles triangle ABC, be produced beyond the base to D, so that $BD=AB$, shew that the square on CD is equal to the square on AB together with twice the square on BC.

IV. 15. Shew how to derive the hexagon from an equilateral triangle inscribed in the circle, and from this construction shew that the side of the hexagon equals the radius of the circle, and that the hexagon is double of the triangle.

1857. I. 35. ABC is an isosceles triangle, of which A is the vertex: AB, AC are bisected in D and E respectively; BE, CD intersect in F: shew that the triangle ADE is equal to three times the triangle DEF.

II. 13. The base of a triangle is given, and is bisected by the centre of a given circle, the circumference of which is the locus of the vertex: prove that the sum of the squares on the two sides of the triangle is invariable.

III. 22. Prove that the sum of the angles in the four segments of the circle, exterior to the quadrilateral, is equal to six right angles.

IV. 4. Circles are inscribed in the two triangles formed by drawing a perpendicular from an angle of a triangle upon the opposite side, and analogous circles are described in relation to the two other like perpendiculars: prove that the

sum of the diameters of the six circles together with the sum of the sides of the original
triangle is equal to twice the sum of the three
perpendiculars.

1858. ɪ. 28. Assuming as an axiom that two straight lines
cannot both be parallel to the same straight
line, deduce Euclid's sixth postulate as a
corollary of the proposition referred to.

ɪɪ. 7. Produce a given straight line, so that the sum
of the squares on the given line and the part
produced may be equal to twice the rectangle
contained by the whole line thus produced and
the produced part.

ɪɪɪ. 19. Describe a circle, which shall touch a given
straight line at a given point and bisect the
circumference of a given circle.

1859. ɪ. 41. Trisect a parallelogram by straight lines drawn
from one of its angular points.

ɪɪ. 13. Prove that, in any quadrilateral, the squares
on the diagonals are together equal to twice
the sum of the squares on the straight lines
joining the middle points of opposite sides.

ɪɪɪ. 31. Two equal circles touch each other externally,
and through the point of contact chords are
drawn, one to each circle, at right angles to
each other: prove that the straight line,
joining the other extremities of these chords,
is equal and parallel to the straight line
joining the centres of the circles.

ɪᴠ. 4. Triangles are constructed on the same base with
equal vertical angles: prove that the locus
of the centres of the escribed circles, each of
which touches one of the sides externally
and the other side and base produced, is an
arc of a circle, the centre of which is on the
circumference of the circle circumscribing the
triangles.

1860.   **L 35.** If a straight line DME be drawn through the
middle point M of the base BC of a triangle
ABC, so as to cut off equal parts AD, AE
from the sides AB, AC, produced if necessary, respectively, then shall BD be equal to
CE.

    II. 14. Shew how to construct a rectangle which shall
be equal to a given square ; (1) when the
sum, and (2) when the difference of two adjacent sides is given.

    III. 36. If two chords AB, AC be drawn from any point
A of a circle, and be produced to D and E,
so that the rectangle AC, AE is equal to the
rectangle AB, AD, then, if O be the centre
of the circle, AO is perpendicular to DE.

    IV. 10. If A be the vertex, and BD the base of the
constructed triangle, D being one of the points
of intersection of the two circles employed in
the construction, and E the other, and AE
be drawn meeting BD produced in F, prove
that FAB is another isosceles triangle of the
same kind.

1861.   **L 32.** If ABC be a triangle, in which C is a right
angle, shew how, by means of Book I., to
draw a straight line parallel to a given
straight line so as to be terminated by CA
and CB and bisected by AB.

    II. 13. If ABC be a triangle, in which C is a right
angle, and DE be drawn from a point D in
AC at right angles to AB, prove, without
using Book III., that the rectangles AB, AE
and AC, AD will be equal.

    III. 32. Two circles intersect in A and B, and CBD is
drawn perpendicular to AB to meet the
circles in C and D ; if AEF bisect either the
interior or exterior angle between CA and
DA, prove that the tangents to the circles at
E and F intersect in a point on AB produced.

1861. IV. 4. Describe a circle touching the side BC of the triangle ABC, and the other two sides produced, and prove that the distance between the points of contact of the side BC with the inscribed circle, and the latter circle, is equal to the difference between the sides AB and AC.

1862. I. 4. Upon the sides AB, BC, and CD of a parallelogram $ABCD$, three equilateral triangles are described, that on BC towards the same parts as the parallelogram, and those on AB, CD towards the opposite parts. Prove that the distances of the vertices of the triangles on AB, CD, from that on BC, are respectively equal to the two diagonals of the parallelogram.

II. 10. Divide a given straight line into two parts, so that the squares on the whole line and on one of the parts may be together double of the square on the other part.

III. 28. A triangle is turned about its vertex, until one of the sides intersecting in that vertex is in the same straight line as the other previously was : prove that the line, joining the vertex with the point of intersection of the two positions of the base, produced if necessary, bisects the angle between these two positions.

IV. 10. Prove that the smaller of the two circles, employed in Euclid's construction, is equal to the circle described about the required triangle.

1863. I. 47. Two triangles ABC, $A'B'C'$ have their sides respectively parallel. BB_1, CC_1 are drawn perpendicular to $B'C'$; CC_2, AA_2 to $C'A'$; and AA_3, BB_3 to $A'B'$. Prove that the sum of the squares on AB_1, BC_2, CA_3 together, is equal to the sum of those on AC_1, BA_2, CB_3 together.

II. 11. Divide a given straight line into two parts, such

that the rectangle contained by the whole and one part may be equal to that contained by the other part and a given straight line.

1863. III. 28. Two equal circles intersect in A, B; PQT perpendicular to AB meets it in T, and the circles in P, Q. AP, BQ meet in R; AQ, BP in S: prove that the angle RTS is bisected by TP.

1864. I. 38. If a quadrilateral figure have two sides parallel, and the parallel sides be bisected, the line joining the points of bisection shall pass through the point in which the diagonals cut one another.

II. 14. Divide a given straight line (when possible) into three parts such that the rectangle contained by two of them shall be equal to a given rectilineal figure, and that the squares on these two parts shall together be equal to the square on the third.

III. 36. If from a given point A without a given circle any two straight lines APQ, ARS, be drawn, making equal angles with the diameter which passes through A, and cutting the circle in P, Q, and R, S, respectively, then PS, QR, shall cut one another in a given point.

IV. 11. If a figure of any odd number of sides have all its angular points on the same circle, and all its angles equal, then shall its sides be equal.

1865. I. 20. Give a geometrical construction for finding a point in a given straight line, the difference of the distances of which from two given points on the same side of the line shall be the greatest possible.

II. 12. The base BC of an isosceles triangle ABC is produced to a point D; AD is joined, and in AD a point E is taken, such that the rectangle AD, AE, is equal to the square on either of the equal sides AB, AC, of the triangle:

prove that the rectangle BD, CD is equal to the rectangle AD, ED.

1865. III. 18. A given straight line is drawn at right angles to the straight line joining the centres of two given circles : prove that the difference between the squares on two tangents drawn, one to each circle, from any point on the given straight line, is constant.

IV. 5. Having given one side of a triangle, and the centre of the circumscribed circle, determine the locus of the centre of the inscribed circle.

1866. I. 33. Prove that a quadrilateral, which has two opposite sides and two opposite obtuse angles equal, is a parallelogram.

Shew that the figure is not necessarily a parallelogram, if the equal angles are acute.

II. 9. Prove this also by superposition of the squares or their halves.

III. 32. If four circles be drawn, each passing through three out of four given points, the angle between the tangents at the intersection of two of the circles is equal to the angle between the tangents at the intersection of the other two circles.

IV. 2. In a given circle inscribe a triangle such that two of the sides of the triangle shall pass through given points and the third side be at a given distance from the centre of the given circle.

1867. I. 16. Any two exterior angles of a triangle are together greater than two right angles.

I. 43. What is the greatest value which these complements, for a given parallelogram, can have?

II. 11. Divide a given straight line into two parts such that the squares on the whole line and on one of the parts shall be together double of the square on the other part.

1867. III. 22. If the chords, which bisect two angles of a triangle inscribed in a circle, be equal, prove that either the angles are equal, or the thiro angle is equal to the angle of an equilateral triangle.

1868. I. 41. *OKBM* and *OLDN* are parallelograms about the diameter of a parallelogram *ABCD*. In *MN*, which is parallel to *BA*, take any point *P* and prove that, if *PC*, produced if necessary, meet *KL* in *Q*, *BP* will be parallel to *DQ*.

II. 12. In a triangle *ABC*, *D*, *E*, *F* are the middle points of the sides *BC*, *CA*, *AB* respectively. and *K*, *L*, *M* are the feet of the perpendiculars on the same sides from the opposite angles. Prove that the greatest of the rectangles contained by *BC* and *DK*, *CA* and *EL*, *AB* and *FM*, is equal to the sum of the other two.

III. 35. Through a point within a circle, draw a chord, such that the rectangle contained by the whole chord and one part may be equal to a given square.

Determine the necessary limits to the magnitude of this square.

IV. 4. If two triangles *ABC*, *A'B'C'* be inscribed in the same circle, so that *AA' BB' CC'* meet in one point *O*, prove that, if *O* be the centre of the inscribed circle of one of the triangles, it will be the centre of the perpendiculars of the other.

1869. I. 40. *ABC* is a triangle, *E* and *F* are two points ; if the sum of the triangles *ABE* and *BCE* be equal to the sum of the triangles *ABF* and *BCF*, then under certain conditions *EF* will be parallel to *AC*. Find these conditions, and determine when the difference instead of the sum of the triangles must be taken.

1869. II. 11. Shew that the point of section lies between the
extremities of the line.

III. 33. An acute-angled triangle is inscribed in a
circle, and the paper is folded along each of
the sides of the triangle : Shew that the
circumferences of the three segments will pass
through the same point. State the equivalent
proposition for an obtuse-angled triangle.

IV. 11. Shew that the circles, each of which touches
two sides of a regular pentagon at the extremities of a third, meet in a point.

1870. I. 26. $ABCD$ is a square and E a point in BC; a
straight line EF is drawn at right angles to
AE, and meets the straight line, which bisects
the angle between CD and BC produced in a
point F : prove that AE is equal to EF.

II. 9. The diagonals of a quadrilateral meet in E, and
F is the middle point of the straight line
joining the middle points of the diagonals :
prove that the sum of the squares on the
straight lines joining E to the angular points
of the quadrilateral is greater than the sum of
the squares on the straight lines joining F to
the same points by four times the square
on EF.

III. 32. AB, CD are parallel diameters of two circles,
and AC cuts the circles in P, Q : prove that
the tangents to the circles at P, Q are parallel.

IV. 10. Hence shew how to describe an equilateral
and equiangular pentagon about a circle without first inscribing one.

1871. I. 38. Through the angular points A, B, C, of a
triangle are drawn three parallel straight lines
meeting the opposite sides in A', B', C' respectively : prove that the triangles $AB'C'$,
$BC'A'$, $CA'B'$ are all equal.

II. 10. Produce a given straight line so that the square
on the whole line thus produced may be
double the square on the part produced.

15

1871. III. 32. The opposite sides of a quadrilateral inscribed in a circle are produced to meet in P, Q, and about the four triangles thus formed circles are described : prove that the tangents to these circles at P and Q form a quadrilateral equal in all respects to the original, and that the line joining the centres of the circles, about the two quadrilaterals, bisects PQ.

IV. 5. A triangle is inscribed in a given circle so as to have its centre of perpendiculars at a given point : prove that the middle points of its sides lie on a fixed circle.

1872. I. 47 If CE, BD be the squares described upon the side AC, and the hypotenuse AB, and if EB, CD intersect in F, prove that AF bisects the angle EFD.

III. 22. Two circles intersect in A, B : PAP', QaQ' are drawn equally inclined to AB to meet the circles in P, P', Q, Q' : prove that PP' is equal to QQ'.

IV. 4. Having given an angular point of a triangle, the circumscribed circle, and the centre of the inscribed circle, construct the triangle.

BOOK V.

SECTION I.

On Multiples and Equimultiples.

Def. I. A GREATER magnitude is a *Multiple* of a less magnitude, when the greater contains the less an exact number of times.

Def. II. A LESS magnitude is a *Sub-multiple* of a greater magnitude, when the less is contained an exact number of times in the greater.

These definitions are applicable not merely to Geometrical magnitudes, such as Lines, Angles, and Triangles ; but also to such as are included in the ordinary sense of the word Magnitude, that is, anything which is made up of parts like itself, such as a Distance, a Weight, or a Sum of Money.

POSTULATE.

Any one magnitude being given, let it be granted that any number of other magnitudes may be found, each of which is equal to the first.

METHOD OF NOTATION.

Let *A* represent a magnitude, not as one of the letters used in Algebra to represent the *measure* of a magnitude, but let *A* stand for the magnitude itself. Thus, if we regard *A* as representing a weight, we mean, not the *number* of pounds contained in the weight, but the weight itself.

Let the words A, B *together* represent the magnitude obtained by putting the magnitude B to the magnitude A.

Let A, A *together* be abbreviated into $2A$,
A, A, A *together* ....$3A$,
and so on.

Let A, A......repeated m times be denoted by mA, m standing for a *whole number.*

Let mA, mA......repeated n times be denoted by nmA, where nm stands for the arithmetical product of the *whole* numbers n and m.

Let $(m+n)A$ stand for the magnitude obtained by putting nA to mA, m and n standing for whole numbers.

These, and these only, are the symbols by which we propose to shorten and simplify the proofs of this Book : capital letters standing, in all cases, for *magnitudes ;* and small letters standing for *whole numbers.*

SCALES OF MULTIPLES.

By taking a number of magnitudes each equal to A, and putting two, three, four......of them together, we obtain a set of magnitudes, depending upon A, and all known when A is known ; namely,

A, $2A$, $3A$, $4A$, $5A$............and so on ;

each being obtained by putting A to the preceding one.

This we call the SCALE OF MULTIPLES of A.

If m be a whole number, mA and mB are called *Equimultiples* of A and B, or, the same multiples of A and B respectively.

AXIOMS.

1. Equimultiples of the same, or of equal magnitudes, are equal to one another.

2. Those magnitudes, of which the same, or equal, magnitudes are equimultiples, are equal to one another.

3. A multiple of a greater magnitude is greater than the same multiple of a less.

4. That magnitude, of which a multiple is greater than the same multiple of another, is greater than that other magnitude.

NOTE 1. If A and B be two commensurable magnitudes, it is easy to show that there is *some* multiple of A, which is equal to *some* multiple of B.

For let M be a common measure of A and B; then the scale of multiples of M is

$$M, 2M, 3M,\ldots\ldots$$

Now *one* of the multiples in this scale, suppose pM, is equal to A, and one ....suppose qM,..............B.

Hence the multiple qpM is equal to qA,   V. Ax. 1.
and the same multiple is equal to pB;
and therefore $qA = pB$.   I. Ax. 1.

PROPOSITION I. (Eucl. v. 1.)

If any number of magnitudes be equimultiples of as many, each of each ; whatever multiple any one of them is of its sub-multiple, the same multiple must all the first magnitudes, taken together, be of all the other, taken together.

Let A be the same multiple of C that B is of D.
Then must A, B together be the same multiple of C, D together that A is of C.

Let $A = C, C, C\ldots\ldots\ldots$repeated m times.
Then $B = D, D, D\ldots\ldots\ldots$repeated m times.

$\therefore A, B$ together $= C, D$; C, D ; C, D;......repeated m times.
$\therefore A, B$ together is the same multiple of C, D together that A is of C.
                                                  Q. E, D.

PROPOSITION II. (Eucl. v. 2.)

If the first be the same multiple of the second that the third is of the fourth, and the fifth the same multiple of the second that the sixth is of the fourth ; the first together with the fifth must be the same multiple of the second, that the third together with the sixth is of the fourth.

Let A, B, C, D, E, F be six magnitudes, such that
A is the same multiple of B, that C is of D, and
E is the same multiple of B, that F is of D.

Then must A, E together be the same multiple of B, that C, F together is of D.

Let $A = B, B, B,$............repeated m times ;
then $C = D, D, D,$............repeated m times.
Also, let $E = B, B, B,$............repeated n times ;
then $F = D, D, D,$............repeated n times.
∴ A, E together $= B, B, B,$............repeated $m+n$ times,
and C, F together $= D, D, D,$............repeated $m+n$ times.
∴ A, E together is the same multiple of B,
that C, F together is of D.

Q. E. D.

PROPOSITION III. (Eucl. v. 3.)

If the first be the same multiple of the second that the third is of the fourth ; and if of the first and third there be taken equimultiples, these must be equimultiples, the one of the second, and the other of the fourth.

Let A be the same multiple of B that C is of D ;
and let E and F be taken equimultiples of A and C.

Then must E and F be equimultiples of B and D.

For let $A = B, B,$............repeated m times$=mB$;
then $C = D, D,$............repeated m times$=mD$.
Again, let $E = A, A,$............repeated n times ;
then $F = C, C,$............repeated n times.
∴ $E = mB, mB,$............repeated n times$=nmB$;
and $F = mD, mD,$............repeated n times$=nmD$.
∴ E is the same multiple of B that F is of D.

Q. E. D.

SECTION II.

On Ratio and Proportion.

Def. III. If A and B be magnitudes of the same kind, the relative greatness of A with respect to B is called the RATIO of A to B.

Note 2. When A and B are *commensurable*, we can estimate their relative greatness by considering what multiples they are of some common standard. But as this method is not applicable when A and B are incommensurable, we have to adopt a more general method, applicable both to commensurable and incommensurable magnitudes.

If A and B be magnitudes of the same kind, commensurable or incommensurable, the scale of multiples of A is

$$A, 2A...mA, (m+1)A...2mA, (2m+1)A...3mA...nmA...$$

and the Ratio of B to A is estimated by considering the position which B, or some multiple of B, occupies among the multiples of A.

If A and B be commensurable, a multiple of B can be found, such that it would occupy *the same place* among the multiples of A, which is occupied by *some one* of the multiples of A; that is, this particular multiple of B represents the same magnitude as that, which is represented by *some one* of the multiples of A. See Note 1, p. 213.

If, for example, the 7th multiple in the scale of B represents the same magnitude as that which is represented by the 5th multiple in the scale of A, or in other words, if $7B = 5A$, we are enabled to form an exact notion of the greatness of B relatively to A.

When A and B are incommensurable, the relation $mA = nB$ can have no existence ; that is, no pair of multiples, one in each of the scales of multiples of A and B, represent the same magnitude. But we can always determine whether a *particular* multiple of B be greater or less than some one of the multiples of A ; that is, we can always find between what two successive multiples of A any given multiple of B lies.

Hence, whether A and B be commensurable or incommensurable, we can always form a *third* scale, in which the multiples of B are distributed among the multiples of A.

Suppose, for example, we discover the following relations between particular multiples of A and B :

B greater than A and less than $2A$,
$2B$ greater than $3A$ and less than $4A$,
$3B$ greater than $5A$ and less than $6A$,

and so on ; the *third* scale will commence thus

$$A, B, 2A, 3A, 2B, 4A, 5A, 3B, 6A,$$

and so on ; the scale not being formed by any law, but constructed by special calculations for each term.

Such a scale we call the SCALE OF RELATION of A and B, and we give the following DEFINITION :—

The Scale of Relation of two magnitudes of the same kind is a list of the multiples of both *ad infinitum*, all arranged in order of magnitude, so that any multiple of either magnitude being assigned, the scale of relation points out between which multiples of the other it lies.

NOTE 3. It may here be remarked that, if A and B be two *finite* magnitudes of the *same* kind, however small B may be, we may, by continuing the scale of multiples of B sufficiently far, at length obtain a multiple of B greater than A.

Also, if B be less than A, *one multiple at least* of the scale of B will lie between each two consecutive multiples of the scale of A. From these considerations we shall be justified in assuming

(1.) That we can always take mB greater than A or than pA.

(2.) That we can always take nB such that it is greater than pA but not greater than qA, provided that B is less than A, and p than q.

We can now make an important addition to Definition III., so that it will run thus :—

If A and B be magnitudes of the same kind, the relative greatness of A with respect to B is called the Ratio of A to B, and this Ratio is determined by, that is, depends solely upon, the order in which the multiples of A and B occur in the Scale of Relation of A and B.

DEF. IV. Magnitudes are said to have a Ratio to each other, which can, being multiplied, exceed each the other.

This definition is inserted to point out that a ratio cannot exist between two magnitudes unless two conditions be fulfilled :— first, the magnitudes must be of the same kind ; secondly, neither of them may be infinitely large or infinitely small.   See Note 3.

DEF. V. When there are four magnitudes, and when any equimultiples of the first and third being taken, and any equimultiples of the second and fourth, if, when the multiple of the first is greater than that of the second, the multiple of the third is greater than that of the fourth, and when the multiple of the first is equal to that of the second, the multiple of the third is equal to that of the fourth, and when the multiple of the first is less than that of the second, the multiple of the third is less than that of the fourth, then the first of the original four magnitudes is said to have to the second the same ratio which the third has to the fourth.

Note 4.—To make Def. v. clearer we give the following illustration. Suppose A, B, C, D to be four magnitudes ; the scales of their multiples will then be—

$$A, 2A, 3A\ldots\ldots\ldots\ldots mA\ldots\ldots\ldots\ldots,$$
$$B, 2B, 3B\ldots\ldots\ldots\ldots nB\ldots\ldots\ldots\ldots,$$
$$C, 2C, 3C\ldots\ldots\ldots\ldots mC\ldots\ldots\ldots\ldots,$$
$$D, 2D, 3D\ldots\ldots\ldots\ldots nD\ldots\ldots\ldots\ldots;$$

where mA, mC stand for *any* equimultiples of A and C, and nB, nD stand for *any* equimultiples of B and D: then the Definition may be stated more briefly thus :

A is said to have the same ratio to B which C has to D, when mA is found in the same position among the multiples of B, in which mC is found among the multiples of D ; or, which is the same thing, *when the order of the multiples of A and B in the Scale of Relation of A and B, is precisely the same as the order of the multiples of C and D in the Scale of Relation of C and D* ; or, when *every* multiple of A is found in the same position among the multiples of B, in which the same multiple of C is found among the multiples of D.

Note 5. The use of Def. v. will be better understood by the following application of it.

To show that rectangles of equal altitude are to one another as their bases.

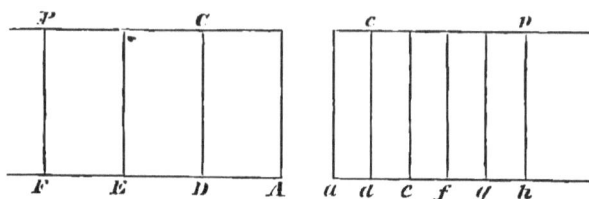

Let AC, ac be two rectangles of equal altitude.

Let B, B' and R, R' stand for the bases and the areas of these rectangles respectively.

Take AD, DE, EF,$\ldots\ldots\ldots\ldots m$ in number, and all equal,
And ad, de, ef, fg, gh,$\ldots\ldots\ldots\ldots n$ in number, and all equal.

Complete the rectangles, as in the diagram.
$$\text{Then base } AF = mB,$$
$$\text{base } ah = nB'.$$
$$\text{rectangle } AP = mR,$$
$$\text{rectangle } ap = nR',$$
Now we can prove, by superposition, that if AF be greater than ah, AP will be greater than ap, and if equal, equal ; and if less, less.

That is, if mB be greater than nB', mR is greater than nR'; and if equal, equal ; and if less, less.

Hence, by Def. v.,
$$B \text{ is to } B' \text{ as } R \text{ is to } R'.$$
Hence we deduce two Corollaries, which are the foundation of the proofs in Book VI.

Cor. I. *Parallelograms of equal altitude are to one another as their bases.*

For the parallelograms are equal to rectangles, on the same bases and between the same parallels.

Cor. II. *Triangles of equal altitude are to one another as their bases.*

For the triangles are equal to the halves of the rectangles, on the same bases and between the same parallels.

N.B.—These Corollaries are proved as a direct Proposition in Eucl. VI. 1.   Cor. II. could not, consistently with Euclid's method, be introduced in this place, for it assumes Proposition XI. of Book V.

DEF. VI. Magnitudes which have the same ratio are called *Proportionals.*

If A, B, C, D be proportionals, it is usually expressed by saying, A is to B as C is to D.

The magnitudes A and C are called the *Antecedents* of the ratios.
.... B and D..................*Consequents*...............

The antecedents are said to be *homologous* to one another, that is, occupying the *same* position in the *ratios* (ὁμόλογοι), and the consequents are said to be homologous to one another.

DEF. VII. When of the equimultiples of four magnitudes, taken as in Def. v., the multiple of the first is greater than [or is equal to] the multiple of the second, but the multiple of the third is not greater than [or is less than] the multiple of the fourth, then the first is said to have to the second a greater ratio, than the third has to the fourth.

NOTE 6. The meaning of Def. VII. may be expressed, after taking the scales of multiples as in the explanation of Def. v., thus :—

A is said to have to B a greater ratio than C has to D, when two whole numbers m and n can be found, such that mA is greater than nB, but mC not greater than nD ; or, such that mA is equal to nB, but mC less than nD.

SECTION III.

Containing the Propositions most frequently referred to in
Book VI.

NOTE 7. The Fifth Book of Euclid may be regarded in two aspects : first, as a Treatise on the Theory of Ratio and Proportion, complete in itself, and depending in no way on the preceding Books of the Elements ; and secondly, as a necessary introduction to the Sixth Book.

If we make the number of references in Book VI. a test of the importance of particular Propositions in Book V., they will be arranged in the following order :—

Proposition V. is referred to 23 times.
  ,,   VI.   ,,   14 ,,
  ,, VIII. ,,   7 ,,
  ,, XXI.   ,,   5 ,,
  ,, XVIII. ,,   3 ,,
  ,, XII.   ,,   2 ,,

Propositions X., XI, XV., XVI., XIX., XXII., are referred to *once.*

It is desirable, then, that the student should observe that the *three* Propositions, which are of especial importance for Book VI., are included in this Section.

PROPOSITION IV.

If four magnitudes be proportionals, and any equimultiples be taken of the first and third, and also any equimultiples of the second and fourth, if the multiple of the first be greater than that of the second, the multiple of the third must be greater than that of the fourth ; and if equal, equal ; and if less, less.

Let A be to B as C is to D,
and let any equimultiples mA, mC be taken of A and C,
and any equimultiples nB, nD............ of B and D.

Then if mA be greater than nB, mC must be greater than nD ; and if equal, equal ; if less, less.

For if mA be greater than nB, but mC not greater than nD, then will A have to B a greater ratio than C has to D ; which is not the case.   V. Def. 7.

Hence if mA be greater than nB, mC must be greater than nD.

Similarly it may be shown that, if mA be equal to, or less than, nB, mC must also be equal to, or less than, nD.

Q. E. D.

N.B.—We have added this Proposition to meet an objection, which might be made to a reference to Definition v., when the *converse* of that Definition is wanted. This reference is of frequent occurrence in Simson's edition.

PROPOSITION V. (Eucl v. 11.)

Ratios that are the same to the same ratio, are the same to one another.
Let A be to B as C is to D,
and E be to F as C is to D.
Then must A be to B as E is to F.

Take of A, C, E any equimultiples mA, mC, mE,
and of B, D, F any equimultiples nB, nD, nF.

Then ∵ A is to B as C is to D,
∴ if mA be greater than nB, mC is greater than nD ;
and if equal, equal ; if less, less.   V. 4.

Again, ∵ C is to D as E is to F,
∴ if mC be greater than nD, mE is greater than nF ;
and if equal, equal ; if less, less.   V. 4.

Hence, if mA be greater than nB, mE is greater than nF ;
and if equal, equal ; if less, less.

. A is to B as E is to F.   V. Def. 5.

Q. E. D.

PROPOSITION VI. (Eucl. v. 7.)

Equal magnitudes have the same ratio to the same magnitude ; and the same has the same ratio to equal magnitudes.

Let A and B be equal magnitudes, and C any other magnitude.

> Then must A be to C as B is to C,
> and C must be to A as C is to B.

Take mA and mB any equimultiples of A and B,
and nC any multiple of C.

Then ∵ $A = B$, ∴ $mA = mB$.   V. Ax. 1.

∴ if mA be greater than nC, mB is greater than nC ;
and if equal, equal ; if less, less.

∴ A is to C as B is to C.   V. Def. 5.

Again, if nC be greater than mA, nC is greater than mB ;
and if equal, equal ; if less, less.

∴ C is to A as C is to B.   V. Def. 5.

Q. E. D.

PROPOSITION VII. (Eucl. v. 8.)

Of two unequal magnitudes, the greater has a greater ratio to any other magnitude than the less has ; and the same magnitude has a greater ratio to the less, of two other magnitudes, than it has to the greater.

Let A and B be any two magnitudes, of which A is the greater, and let D be any other magnitude.

Then must the ratio of A to D be greater than the ratio of B to D.

Take such equimultiples of A and B, qA and qB, that each of them may be greater than D.   Note 3, p. 216.

Then ∵ A is greater than B,
∴ qA is greater than qB.   V. Ax. 3.
Let $qA = qB$, R together.

Then, however small R may be, we can find a multiple of R, suppose mR, such that mR is greater than qB.   Note 3.

Take equimultiples of qA and qB, mqA and mqB, and take a multiple of D, nD, such that nD is not less than mqB and not greater than $(mq+q)B$.   Note 3.

Then ∵ $mqA = mqB$, mR together,   V. 1.
and mR is greater than qB,
∴ mqA is greater than $(mq+q)B$,
and, *a fortiori*, mqA is greater than nD.
But mqB is not greater than nD,
∴ the ratio of A to D is greater than the ratio of B to D.
V. Def. 7.

Also, the ratio of D to B must be greater than the ratio of D to A.

For, the same multiples being taken as before,
∵ nD is not less than mqB,
and nD is less than mqA,
∴ D has to B a greater ratio than D has to A.
V. Def. 7.

Q. E. D.

PROPOSITION VIII. (Eucl. v. 9.)

Magnitudes, which have the same ratio to the same magnitude, are equal to one another ; and those, to which the same magnitude has the same ratio, are equal to one another.

Let A and B have the same ratio to C.

Then must $A = B$.

For if A were greater than B,

A would have a greater ratio to C than B has to C; V. 7. which is not the case.

And if A were less than B,

B would have a greater ratio to C than A has to C; V. 7 which is not the case.

$$\therefore A = B.$$

Next, let C have the same ratio to A that C has to B.

Then must $A = B$.

For we can show, as before, that A cannot be greater or le · than B.

$$\therefore A = B. \qquad \text{Q. E. D.}$$

PROPOSITION IX. (Eucl. v. 10.)

That magnitude, which has a greater ratio than another has to the same magnitude, is the greater of the two; and that magnitude, to which the same has a greater ratio than it has to another magnitude, is the less of the two.

Let A have to C a greater ratio than B has to C.

Then must A be greater than B.

For if A were equal to B, then would A have the same ratio to C that B has to C; which is not the case. V. 8.

And if A were less than B, then would A have to C a ratio less than that which B has to C; which is not the case. V. 7.

$$\therefore A \text{ is greater than } B.$$

Next, let C have a greater ratio to B than it has to A.

Then must B be less than A.

For if B were equal to A, then would C have the same ratio to B which it has to A ; which is not the case. V. 8.

And if B were greater than A, then C would have to B a ratio less than that which C has to A ; which is not the case. V. 7.

$$\therefore B \text{ is less than } A. \qquad \text{Q. E. D.}$$

Proposition X. (Eucl. v. 12.)

If any number of magnitudes be proportionals, as one of the antecedents is to its consequent, so must all the antecedents taken together be to all the consequents.

Let any number of magnitudes A, B, C, D, E, F...be proportionals,
that is, A to B as C to D and as E is to F...

Then must A be to B as A, C, E...together is to B, D, F...together.

Take of $A, C, E,$...any equimultiples mA, mC, mE...
    and of B, D, F...any equimultiples nB, nD, nF...

Then $\because$ A is to B as C is to D and as E is to F...

$\therefore$ if mA be greater than nB, mC is greater than nD,
and mE is greater than nF...; and if equal, equal; if less,
less.   V. 4.

$\therefore$ if mA be greater than nB, mA, mC, mE...together are
greater than nB, nD, nF...together; and if equal, equal; if
less, less.

Now mA and mA, mC, mE...together are equimultiples of
A and A, C, E...together.   V. 1.

And nB and nB, nD, nF...together are equimultiples of
B and B, D, F...together.

$\therefore$ A is to B as A, C, E...together is to B, D, F...together.
        V. Def. 5.
        Q. E. D.

Proposition XI. (Eucl. v. 15.)

Magnitudes have the same ratio to one another which their equimultiples have.

Let A be the same multiple of C that B is of D.

Then must C be to D as A to B.

Divide A into magnitudes E, F, G,...each equal to C,
    and B into magnitudes H, K, L,... each equal to D,
the number of the magnitudes being the same in both cases,
because A and B are *equimultiples* of C and D.

Then $\because$ E, F, G.........are all equal,
    and H, K, L.........are all equal.

$\therefore$ E is to H, as F to K, as G to L...   V. 6
$\therefore$ E is to H as E, F, G...together is to H, K, L..
together,   V. 10

    that is, E is to H as A to B;
    and $\because$ $E = C$, and $H = D$,
    $\therefore$ C is to D as A to B.   Q. E. D.

SECTION IV.

On Proportion by Inversion, Alternation, and Separation.

PROPOSITION XII. (Eucl. v. B.)

If four magnitudes be proportionals, they must also be proportionals when taken inversely.

Let A be to B as C is to D.

Then inversely B must be to A as D is to C.

Take of A and C any equimultiples mA and mC, and of B and D any equimultiples nB and nD.

Then ∵ A is to B as C is to D,

∴ if mA be greater than nB, mC is greater than nD; and if equal, equal; if less, less.   V. 4.

Hence, if nB be greater than mA, nD is greater than mC; and if equal, equal; if less, less.

∴ B is to A as D is to C.   V. Def. 5.

Q. E. D.

PROPOSITION XIII. (Eucl. v. 13.)

If the first has to the second the same ratio which the third has to the fourth, but the third to the fourth a greater ratio than the fifth has to the sixth ; the first must also have to the second a greater ratio than the fifth has to the sixth.

Let A have to B the same ratio that C has to D, but C to D a greater ratio than E has to F.

Then must A have to B a greater ratio than E has to F.

For ∵ C has to D a greater ratio than E has to F, we can find such equimultiples of C and E, suppose mC and mE, and such equimultiples of D and F, suppose nD and nF, that mC is greater than nD, but mE not greater than nF.   V. Def. 7.

Then ∵ A is to B as C is to D,   Hyp.
and mC is greater than nD,
∴ mA is greater than nB.   V. 4.

And mE is not greater than nF.

∴ A has to B a greater ratio than E has to F.   V. Def. 7.

Q. E. D.

PROPOSITION XIV. (Eucl. v. 14.)

If the first has to the second the same ratio which the third has to the fourth ; then, if the first be greater than the third the second must be greater than the fourth ; and if equal, equal ; and if less, less.

Let A have the same ratio to B that C has to D.

Then if A be greater than C, B must be greater than D.

For ∵ A is greater than C,
and B is any other magnitude,
∴ A has a greater ratio to B than C has to B.   V. 7.

But A is to B as C is to D.

∴ C has a greater ratio to D, than C has to B.   V. 13.

∴ B is greater than D.   V. 9.

Similarly it may be shown that if A be less than C, B must be less than D ; and that if A be equal to C, B must be equal to D.   Q. E. D.

PROPOSITION XV. (Eucl. v. 16.)

If four magnitudes of the same kind be proportionals, they must also be proportionals when taken alternately.

Let A, B, C, D be four magnitudes of the same kind, and let A be to B as C is to D.

Then alternately A must be to C as B is to D.

Take of A and B any equimultiples mA and mB, and of C and D any equimultiples nC and nD.

Then ∵ mA is to mB as A is to B,   V. 11.
and C is to D as A is to B,   Hyp.

∴ mA is to mB as C is to D.   V. 5.

But nC is to nD as C is to D ;   V. 11.
and ∴ mA is to mB as nC is to nD.   V. 5.

If ∴ mA be greater than nC, mB is greater than nD ; and if equal, equal ; if less, less.   V. 14.

∴ A is to C as B is to D.   V. Def. 5.
    Q. E. D.

PROPOSITION XVI. (Eucl. v. 18.)

If magnitudes taken separately be proportionals, they must be proportionals also when taken jointly.

Let A have the same ratio to B that C has to D.

Then must A, B together have the same ratio to B, that C, D together has to D.

First, when all the magnitudes are of the same kind,

∵ A is to B as C is to D,

∴ A is to C as B is to D. $\mathbf{5}$. 15.

∴ A, B together is to C, D together as B is to D, $\mathbf{V}$. 10.

and ∴ A, B together is to B as C, D together is to D. $\mathbf{V}$. 15.

Next, when all the magnitudes are not of the same kind, we may employ a method of proof which includes the former case : thus—

Take of A, B, C, D any equimultiples mA, mB, mC, mD, and of B and D take any equimultiples nB, nD.

Then ∵ A is to B as C is to D,

∴ if mA be greater than nB, mC is greater than nD ; and if equal, equal ; if less, less.   V. 4.

If then mA, mB together be greater than mB, nB together,

mC, mD together is greater than mC, nD together ;

and if equal, equal ; if less, less.   I. Ax. 2, 4.

Now mA, mB together is the same multiple of A, B together that mC, mD together is of C, D together ;   V. 1.

and mB, nB together is the same multiple of B that mD, nD together is of D.   V. 2.

∴ A, B together is to B as C, D together is to D.   V. Def. 5.

Q. E. D

SECTION V.

Containing the Propositions occasionally referred to in
Book VI.

PROPOSITION XVII. (Eucl. v. 4.)

If the first of four magnitudes has to the second the same ratio which the third has to the fourth, and any equimultiples of the first and third be taken, and also any equimultiples of the second and fourth, then must the multiple of the first have the same ratio to the multiple of the second which the multiple of the third has to that of the fourth.

If A be to B as C is to D,
   and mA, mC be taken equimultiples of A and C,
   and nB, nD.............................of B and D,
then must mA *be to* nB *as* mC *is to* nD.

Take of mA, mC any equimultiples pmA, pmC,
and of nB, nD.......................qnB, qnD.

Then pmA, pmC are equimultiples of A and C,   **V. 3.**
and qnB, qnD.......................of B and D.   **V. 3.**

And ∵ A is to B as C is to D,
   ∴ if pmA be greater than qnB,
      pmC is greater than qnD ;   **V. 4.**
and if equal, equal ; if less, less.

Then ∵ pmA, pmC are equimultiples of mA, mC,
   and qnB, qnD.......................of nB, nD,
   ∴ mA is to nB as mC is to nD.   **V. Def. 5.**

Q. E. D.

PROPOSITION XVIII. (Eucl. v. A.)

If the first of four magnitudes have the same ratio to the second that the third has to the fourth, then, if the first be greater than the second, the third must be greater than the fourth ; and if equal, equal ; and if less, less.

Let A be to B as C is to D.

Then if A be greater than B, C must be greater than D ; and if equal, equal ; and if less, less.

Take any equimultiples of each, mA, mB, mC, mD.

Then $\because$ A is to B as C is to D,

$\therefore$ if mA be greater than mB, mC is greater than mD ;

and if equal, equal ; and if less, less.   V. 4.

    First, suppose A *greater* than B,

        then mA is greater than mB,   V. Ax. 3.

        and $\therefore$ mC is greater than mD,

        and $\therefore$ C is greater than D.   V. Ax. 4.

Similarly the other cases may be proved.

Q. E. D.

PROPOSITION XIX. (Eucl. v. D.)

If the first be to the second as the third is to the fourth, and if the first be a multiple, or a submultiple, of the second, the third must be the same multiple, or the same submultiple, of the fourth.

Let A be to B as C is to D,

and, first, let A be a *multiple* of B.

    Then must C be the same multiple of D.

Let $A = mB$, and take mD the same multiple of D that A is of B.

Then $\because$ A is to B as C is to D,

    $\therefore$ A is to mB as C is to mD.   V. 17.

But $A = mB$, and $\therefore$ $C = mD$.   V. 18.

Next, let A be a *submultiple* of B.
Then must C be the same submultiple of D.

For ∵ A is to B as C is to D,
∴ B is to A as D is to C,   **V. 12.**

Now B is a multiple of A,
and ∴ D is the same multiple of C, by the first case.
Hence C is the same submultiple of D, that A is of B.

Q. E. D.

PROPOSITION XX. (Eucl. v. 20.)

If there be three magnitudes, and other three, which have the same ratio, taken two and two, then, if the first be greater than the third, the fourth must be greater than the sixth ; and if equal, equal ; if less, less.

Let A, B, C be three magnitudes, and D, E, F other three,
and let A be to B as D is to E,
and B be to C as E is to F.
Then if A be greater than C, D must be greater than F ; and if equal, equal ; if less, less.

First, if A be greater than C,
A has to B a greater ratio than C has to B.   **V. 7.**

But C has to B the same ratio that F has to E, **Hyp. & V. 12.**
∴ A has to B a greater ratio than F has to E.

∴ D has to E a greater ratio than F has to E.   **V. 13.**
∴ D is greater than F.   **V. 9.**

Similarly the other cases may be proved.

Q. E. D.

PROPOSITION XXI. (Eucl. v. 22.)

If there be any number of magnitudes, and as many others, which have the same ratio taken two and two in order, the first must have to the last of the first magnitudes the same ratio which the first of the others has to the last of these.

First, let there be three magnitudes *A*, *B*, *C*, and other three *D*, *E*, *F*.

And let *A* be to *B* as *D* is to *E*,
and *B* be to *C* as *E* is to *F*.

Then must A be to C as D is to F.

Take of *A* and *D* any equimultiples *mA*, *mD*,
of *B* and *E*........................*nB*, *nE*,
of *C* and *F*........................*pC*, *pF*.

Then ∵ *A* is to *B* as *D* is to *E*,
∴ *mA* is to *nB* as *mD* is to *nE*.   **V. 17.**

So also, *nB* is to *pC* as *nE* is to *pF*.

∴ if *mA* be greater than *pC*, *mD* is greater than *pF*, and if equal, equal ; if less, less.   V. 20.

∴ *A* is to *C* as *D* is to *F*.   V. Def. 5.

The proposition may be easily extended to any number of magnitudes.   Q. E. D.

PROPOSITION XXII. (Eucl. v. 24.)

If the first have to the second the same ratio which the third has to the fourth, and the fifth have to the second the same ratio which the sixth has to the fourth, then the first and fifth together must have to the second the same ratio which the third and sixth together have to the fourth.

Let *A* be to *B* as *C* is to *D*,
and *E* be to *B* as *F* is to *D*.

Then must A, E together be to B as C, F together is to D

For ∵ *E* is to *B* as *F* is to *D*,
∴ *B* is to *E* as *D* is to *F*.   V. 12.

And ∵ *A* is to *B* as *C* is to *D*,
and *B* is to *E* as *D* is to *F*,
∴ *A* is to *E* as *C* is to *F*.   V. 21.

∴ *A*, *E* together is to *E* as *C*, *F* together is to *F*,   V. 16.

and *E* is to *B* as *F* is to *D* ;

∴ *A*, *E* together is to *B* as *C*, *F* together is to *D*.   V. 21.

         Q. E. D.

SECTION VI.

Containing the Propositions to which no reference is made in Book VI.

PROPOSITION XXIII. (Eucl. v. 5.)

If one magnitude be the same multiple of another, which a magnitude taken from the first is of a magnitude taken from the other, the remainder must be the same multiple of the remainder, that the whole is of the whole.

Let B and D be the magnitudes which are taken away, and A and C the magnitudes which remain, then A, B together, and C, D together will be the wholes.

And let A, B together be the same multiple of C, D together, that B is of D.

Then must A be the same multiple of C that A, B together is of C, D together.

Take E the same multiple of C that B is of D,

Then E, B together is the same multiple of C, D together that B is of D.   V. 1.

But A, B together is the same multiple of C, D together that B is of D.

∴ E, B together $= A$, B together,   V. Ax. 1.
and ∴ $E = A$.   I. Ax. 3.

∴ A is the same multiple of C that B is of D.

                                           Q. E. D.

If two magnitudes be equimultiples of two others, and if equimultiples of these be taken from the first two, the remainders are either equal to these others, or equimultiples of them.

Let B and D be the magnitudes which are taken away,
and A and C the magnitudes which remain ;
then A, B together and C, D together will be the wholes.

Let A, B together be the same multiple of P,
that C, D together is of Q,
and let B be the same multiple of P, that D is of Q.

*Then must A and C be equal respectively to P and Q,
or A and C be equimultiples of P and Q.*

For let A, B together $= P$, P.........repeated $m + n$ times,
then C, D together $= Q$, Q.........repeated $m + n$ times.

Also, let $B = P$, P.........repeated n times,
then $D = Q$, Q.........repeated n times.

Hence $A = P$, P.........repeated m times,
and $C = Q$, Q.........repeated m times.

If then $A = P$, $m = 1$, and $\therefore C = Q$;
and if A be a multiple of P, C is the same multiple of Q.

Q. E. D.

PROPOSITION XXV. (Eucl. v. 17.)

If magnitudes, taken jointly, be proportionals, they shall also be proportionals when taken separately; that is, if two magnitudes together have to one of them the same ratio which two others have to one of these, the remaining one of the first two must have to the other the same ratio which the remaining one of the last two has to the other of these.

Let A, B together have the same ratio to B that C, D together have to D.

Then must A be to B as C to D.

Take of A, B, C, D any equimultiples mA, mB, mC, mD, and again of B, D take any equimultiples nB, nD.

Then ∵ mA is the same multiple of A that mB is of B,
∴ mA, mB together is the same multiple of A, B together that mA is of A.   V. 1.

And ∵ mC is the same multiple of C that mD is of D,
∴ mC, mD together is the same multiple of C, D together that mC is of C.   V. 1.

But mA is the same multiple of A that mC is of C.
∴ mA, mB together is the same multiple of A, B together that mC, mD together is of C, D together.

Again, mB, nB together is the same multiple of B that mD, nD together is of D.

Now, since A, B together is to B as C, D together is to D,
∴ if mA, mB together be greater than mB, nB together, mC, mD together is greater than mD, nD together; and if equal, equal; if less, less.   V. 4.

That is, if mA be greater than nB, mC is greater than nD; and if equal, equal; if less, less.   I. Ax. 3, 5.

∴ A is to B as C is to D.   V. Def. 5.

Q. E. D.

PROPOSITION XXVI. (Eucl. v. 19.)

If a whole magnitude be to a whole as a magnitude taken from the first is to a magnitude taken from the other, the remainder must be to the remainder as the whole is to the whole.

Let A, B together have the same ratio to C, D together that B has to D.

Then must A be to C as A, B together is to C, D together.

For ∵ A, B together is to C, D together as B is to D,

∴ A, B together is to B as C, D together is to D,   V. 15

and ∴ A is to B as C is to D,   V. 25,

Hence A is to C as B is to D.   V. 15.

But A, B together is to C, D together as B is to D.   Hyp.

∴ A is to C as A, B together is to C, D together.   V. 5.

Q. E. D.

PROPOSITION XXVII. (Eucl. v. 21.)

If there be three magnitudes, and other three, which have the same ratio, taken two and two, but in a cross order, then if the first be greater than the third, the fourth must be greater than the sixth ; and if equal, equal ; and if less, less.

Let A, B, C be three magnitudes, and D, E, F other three,
  and let A be to B as E is to F,
  and B be to C as D is to E.

*Then if A be greater than C, D must be greater than F;
and if equal, equal ; and if less, less.*

First, if A be *greater* than C,

  A has to B a greater ratio than C has to B,   V. 7.

and ∴ E has to F a greater ratio than C has to B.   V. 13.

Now ∵ B is to C as D is to E,   --yp.

∴ C is to B as E is to D.   V. 12

Hence E has to F a greater ratio than E has to D.

∴ D is greater than F.   V. 9.

Similarly the other cases may be proved.

Q. E. D.

PROPOSITION XXVIII. (Eucl. **v. 23.**)

If there be any number of magnitudes, and as many others, which have the same ratio, taken two and two in a cross order, the first must have to the last of the first magnitudes the same ratio which the first of the others has to the last of these.

Let *A, B, C* be three magnitudes, and *D, E, F* other three,
and let *A* be to *B* as *E* is to *F*,
and *B* be to *C* as *D* is to *E*.
Then must A be to C as D is to F.

Of *A, B, D* take any equimultiples *mA, mB, mD*, and of *C, E, F* take any equimultiples *nC, nE, nF*.

Now ∵   *A* is to   *B* as   *E* is to   *F*,
∴ *mA* is to *mB* as *nE* is to *nF* ;   V. 11, and V. 5.

and ∵   *B* is to   *C* as   *D* is to   *E*,
∴ *mB* is to *nC* as *mD* is to *nE*.   V. 17.

Hence, if *mA* be greater than *nC, mD* is greater than *nF*, and if equal, equal ; and if less, less.   V. 27.
∴ *A* is to *C* as *D* is to *F*.   V. Def. 5.

The proposition may be easily extended to any number of magnitudes.

Q. E. D.

If four magnitudes of the same kind be proportionals, the greatest and least of them together must be greater than the other two together.

Let A be to B as C is to D,
and let A be the greatest of the four magnitudes, and consequently D the least.   V. 18, and V. 14.
Then must A, D together be greater than B, C together.

Let $A = B$, P together, and $C = D$, Q together.
   Then ∵ B, P together is to B as D, Q together is to D,
       ∴ P is to B as Q is to D,   V. 25.
and B is greater than D.
       ∴ P is greater than Q.   V. 14.

   Hence P, B, D together are greater than Q, B, D
together.   I. Ax. 4.
     ∴ A, D together are greater than B, C together.

                                     Q. E. D.

PROPOSITION XXX. (Eucl. v. C.)

If the first be the same multiple of the second, or the same submultiple of it, that the third is of the fourth, the first must be to the second as the third is to the fourth.

First, let A be the same *multiple* of B, that C is of D.
Then must A be to B as C is to D.

Let $A = pB$ and $\therefore C = pD$.

Take of A and C any equimultiples mA, mC, and of B and D any equimultiples nB, nD.

Then $mA = mpB$ and $mC = mpD$.　　　　V. 3.

Now if mpB be greater than nB,
　　　　mpD is greater than nD ;
and if equal, equal ; if less, less.

That is, if mA be greater than nB, mC is greater than nD ; and if equal, equal ; and if less, less.

　　$\therefore A$ is to B as C is to D.　　　　V. Def. 5.

Next, let A be the same *submultiple* of B, that C is of D.
Then must A be to B as C is to D.

For $\because A$ is the same submultiple of B, that C is of D,
　$\therefore B$ is the same multiple of A, that D is of C,
　$\therefore B$ is to A as D is to C, by the first case,

and $\therefore A$ is to B as C is to D.　　　　V. 12.

QED.

17

PROPOSITION XXXI. (Eucl. v. E.)

If four magnitudes be proportionals, they must also be proportionals by conversion; that is, the first must be to its excess above the second as the third is to its excess above the fourth.

Let A, B together be to B as C, D together is to D.

Then must A, B together be to A as C, D together is to C.

For ∵ A, B together is to B as C, D together is to D,

∴ A is to B as C is to D,   V. 25.

and ∴ B is to A as D is to C,   V. 12.

and ∴ A, B together is to A as C, D together is to C.   V. 16.

Q. E. D.

BOOK VI.

The chief subject of this Book is the Similarity of Rectilinear Figures.

DEF. I. Two rectilinear figures are called *similar*, when they satisfy two conditions :—

I. For every angle in one of the figures there must be a corresponding equal angle in the other.

II. The sides containing any one of the angles in one of the figures must be in the same ratio as the sides containing the corresponding angle in the other figure: the antecedents of the ratios being sides which are adjacent to equal angles in each figure.

Thus ABC and DEF are similar triangles, if the angles at A, B, C be equal to the angles at D, E, F, respectively, and

if BA be to AC as ED is to DF,

and AC be to CB as DF is to FE,

and CB be to BA as FE is to ED.

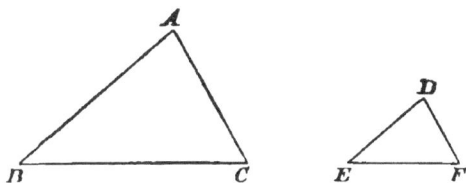

The sides adjacent to equal angles in the triangles are thus *homologous*, that is, BA, AC, CB are respectively homologous to ED, DF, FE.

It will be shown in Prop. IV. that in the case of triangles the second of the above conditions follows from the first.

In the case of quadrilaterals and polygons *both* conditions are necessary : thus any two rectangles have each angle of the one equal to each angle of the other, but they are not necessarily similar figures.

$N.B.$—The very important Prop. XXV. (Eucl. VI. 33) is independent of all the other Propositions in this Book, and might be placed with advantage at the very commencement of the Book.

PROPOSITION I.   THEOREM.

Triangles of the same altitude are to one another as their bases.

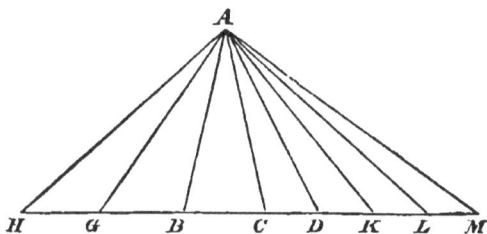

Let the △s *ABC*, *ADC* have the same altitude, that is, the perpendicular drawn from *A* to *BD*.

Then must △ ABC be to △ ADC as base BC is to base DC.

In *DB* produced take any number of straight lines *BG*, *GH* each = *BC*.   I. 3.

In *BD* produced take any number of straight lines *DK*, *KL*, *LM* each = *DC*.   I. 3.

Join *AG*, *AH* ; *AK*, *AL*, *AM*.

Then ∵ *CB*, *BG*, *GH* are all equal,

∴ △s *ABC*, *AGB*, *AHG* are all equal.   I. 38.

∴ △ *AHC* is the same multiple of △ *ABC* that *HC* is of *BC*.

So also,

△ *AMC* is the same multiple of △ *ADC* that *MC* is of *DC*.

And △ *AHC* is equal to, greater than, or less than △ *AMC*, according as base *HC* is equal to, greater than, or less than base *MC*.   I. 38.

Now △ *AHC* and base *HC* are equimultiples of △ *ABC* and base *BC*,

and △ *AMC* and base *MC* are equimultiples of △ *ADC* and base *DC*.

∴ △ *ABC* is to △ *ADC* as base *BC* is to base *DC*.   V. Def. 5.

Q. E. D.

Cor. I. *Parallelograms of the same altitude are to one another as their bases.*

Let $ACBE$, $ACDF$ be parallelograms having the same altitude, that is, the perpendicular drawn from A to BD.

Then must $\square ACBE$ be to $\square ACDF$ as BC is to DC.

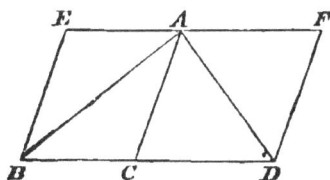

For $\square ACBE$ = twice $\triangle ABC$,   I. 41.
and $\square ACDF$ = twice $\triangle ADC$.   I. 41.

$\therefore \square ACBE$ is to $\square ACDF$ as $\triangle ABC$ is to $\triangle ADC$,   V. 11.
and $\therefore \square ACBE$ is to $\square ACDF$ as BC is to DC.   V. 5.

Q. E. D.

Cor. II. *Triangles and Parallelograms, that have* EQUAL *altitudes, are to one another as their bases.*

Let the figures be placed, so as to have their bases in the same straight line ; and having drawn perpendiculars from the vertices of the triangles to the bases, the straight line, which joins the vertices, is parallel to that, in which their bases are, because the perpendiculars are both equal and parallel to one another.   I. 33.

Then, if the same construction be made as in the Proposition, the demonstration will be the same.

Ex. 1. ABC, DEF are two parallel straight lines ; show that the triangle ADE is to the triangle FBC as DE is to BC.

Ex. 2. If, from any point in a diagonal of a parallelogram, straight lines be drawn to the extremities of the other diagonal, the four triangles, into which the parallelogram is then divided, must be equal, two and two.

PROPOSITION II. THEOREM.

If a straight line be drawn parallel to one of the sides of a triangle, it must cut the other sides, or those sides produced, proportionally.

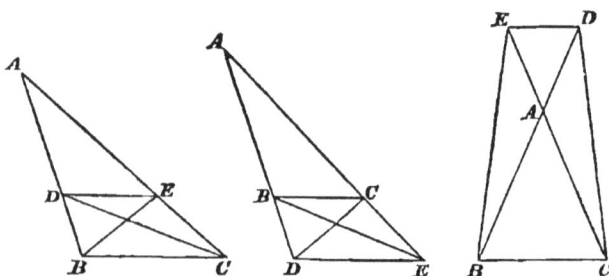

Let DE be drawn ‖ to BC, a side of the $\triangle$ ABC.

Then must BD be to DA as CE to EA.

Join BE, CD.

Then ∵ $\triangle$ $BDE = \triangle$ CDE, on the same base DE
and between the same ‖s, DE, BC.　　　　　　　　　　I. 37.

∴ $\triangle$ BDE is to $\triangle$ ADE as $\triangle$ CDE is to $\triangle$ ADE　　V. 6.

But $\triangle$ BDE is to $\triangle$ ADE 　as　 BD 　is to　 DA,　　VI. 1.

and $\triangle$ CDE is to $\triangle$ ADE 　as　 CE 　is to　 EA ;　VI. 1.

∴　 BD 　is to　 DA 　as　 CE 　is to　 EA.　　V. 5.

Ex. 1. If any two straight lines be cut by three parallel l̦nes, they are cut proportionally. (*N.B.*—This is of great use.)

Ex. 2. If two sides of a quadrilateral be parallel to each other, a straight line, drawn parallel to either of them, shall cut the other sides, or these produced, proportionally.

And Conversely,

If the sides, or the sides produced, be cut proportionally, the straight line which joins the points of section must be parallel to the remaining side of the triangle.

Let the sides AB, AC of the $\triangle ABC$, or these produced, be cut proportionally in D and E, so that
$$BD \text{ is to } DA \text{ as } CE \text{ is to } EA,$$
and join DE.
Then must DE be parallel to BC.

The same construction being made,
∵ BD is to DA as CE is to EA,
and BD is to DA as $\triangle BDE$ is to $\triangle ADE$,   VI. 1.
and CE is to EA as $\triangle CDE$ is to $\triangle ADE$,   VI. 1.
∴ $\triangle BDE$ is to $\triangle ADE$ as $\triangle CDE$ is to $\triangle ADE$, V. 5.
and ∴ $\triangle BDE = \triangle CDE$;   V. 8.
and they are on the same base DE ;
∴ DE is ∥ to BC.   I. 39.

Q. E. D.

Ex. 3. If there be four parallel straight lines, two of these lines intercept upon two given lines, of unlimited length, OA, OB, parts proportional to the parts intercepted upon OA, OB, by the remaining two parallel straight lines.

Ex. 4. If the four sides of a quadrilateral figure be bisected, the lines joining the points of bisection will form a parallelogram.

Ex. 5. A quadrilateral figure has two parallel sides : shew that the straight line, joining the point of intersection of its other two sides produced and the point of intersection of its diagonals, bisects the two parallel sides.

PROPOSITION III. THEOREM.

If the vertical angle of a triangle be bisected by a straight line, which also cuts the base, the segments of the base must have the same ratio, which the other sides of the triangle have to one another.

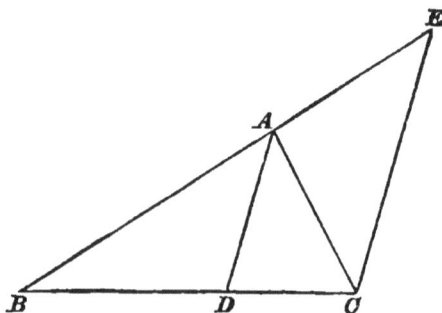

Let $\angle BAC$ of $\triangle$ ABC be bisected by the st. line AD, which meets the base in D.

Then must BD be to DC as BA is to AC.

Through C draw $CE \parallel$ to DA,   I. 31.

and let BA produced meet CE in E.

Then $\angle BAD$=interior $\angle AEC$,   I. 29.

and $\angle CAD$ =alternate $\angle ACE$,   I. 29.

But $\angle BAD = \angle CAD$, by hypothesis,

and $\therefore \angle AEC = \angle ACE$,   Ax. I.

and $\therefore$ AC = AE.   I. B. Cor.

Then $\because AD$ is $\parallel$ to EC, a side of $\triangle$ BEC,

$\therefore BD$ is to DC as BA is to AE,   VI. 2.

and $\therefore BD$ is to DC as BA is to AC.   V. 6.

Ex. 1. Shew that in a parallelogram the diagonals do not bisect the angles, unless the sides are equal.

Ex. 2. Shew how to trisect a straight line of finite length.

Ex. 3. Shew that the bisectors of the angles of a triangle meet in the same point.

Ex. 4. The bisectors of the angles A and C, of a triangle ABC, meet the opposite sides in the points D and F: BA and BC are produced to F' and D', so that AF', AC and CD' are all equal: prove that $F'D'$ is parallel to FD.

And Conversely,

If the segments of the base have the same ratio, which the other sides of the triangles have to one another, the straight line, drawn from the vertex to the point of section, must bisect the vertical angle.

Let BD be to DC as BA is to AC,

and join AD.

Then must $\angle BAD = \angle CAD$.

The same construction being made,

$\because BD$ is to DC as BA is to AC,	Hyp.
and BD is to DC as BA is to AE,	VI. 2.
$\therefore BA$ is to AC as BA is to AE,	V. 5.
and $\therefore AC = AE$,	V. 8.
and $\therefore \angle AEC = \angle ACE$.	I. A.
But $\angle AEC$ = exterior $\angle BAD$,	I. 29.
and $\angle ACE$ = alternate $\angle CAD$,	I. 29.
$\therefore \angle BAD = \angle CAD$.	Ax. 1.

Q. E. D.

Ex. 5. Two straight lines are drawn, bisecting the angles at the base of an isosceles triangle. Shew that the straight line, joining the points, in which they cut the sides, is parallel to the base.

PROPOSITION A. THEOREM.

If the exterior angle of a triangle be bisected by a straight line, which also cuts the base produced, the segments, between the dividing straight line and the extremities of the base, must have the same ratio, which the other sides of the triangle have to one another.

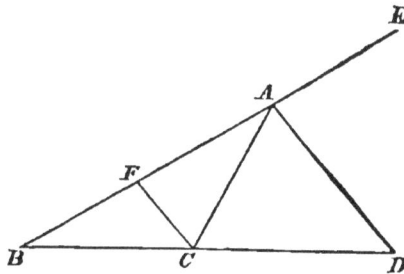

Let ∠ *EAC*, an ext ^{r} ∠ of the △ *ABC*, be bisected by the st. line *AD* which meets the base produced in *D*.

Then must BD be to DC as BA is to AC.

Through *C* draw *CF* ∥ to *DA*, meeting *AB* in *F*. I. 31.

Then ∠ *EAD* = interior ∠ *AFC*, I. 29.

and ∠ *CAD* = alternate ∠ *ACF*. I. 29.

But ∠ *EAD* = ∠ *CAD*, by hypothesis.

∴ ∠ *AFC* = ∠ *ACF*, Ax. 1.

and ∴ *AC* = *AF*. I. B. Cor.

Then ∵ *AD* is ∥ to *FC*, a side of △ *FBC*,

∴ *BD* is to *DC* as *BA* is to *AF*, VI. 2.

and ∴ *BD* is to *DC* as *BA* is to *AC*. V. 6.

Ex. 1. If the angles at the base of the triangle be equal, how is the proposition modified ?

Ex. 2. If *B* be any point in a straight line *AC*, intersected by another, *CD*, give a geometrical construction for determining a point *D* in *CD*, such that *AD* is to *DB* as *AC* is to *CB*.

If the segments of the base produced have the same ratio, which the other sides of the triangle have to one another, the straight line drawn from the vertex to the point of section must bisect the exterior angle of the triangle.

Let BD be to DC as BA is to AC,

and join AD.

Then must $\angle CAD = \angle EAD$.

For, the same construction being made,

$\therefore BD$ is to DC as BA is to AC,	Hyp.
and BD is to DC as BA is to AF,	VI. 2.
$\therefore BA$ is to AC as BA is to AF,	V. 5.
and $\therefore AC = AF$,	V. 8.
and $\therefore \angle AFC = \angle ACF$.	I. A.
But $\angle AFC =$ exterior $\angle EAD$,	I. 29.
and $\angle ACF =$ alternate $\angle CAD$,	I. 29.
and $\therefore \angle CAD = \angle EAD$.	Ax.1.

Q. E. D.

Ex. 3. If the base be divided into two segments, having the same ratio with the segments specified in the Proposition, the straight lines, drawn from the two points of section to the vertex of the triangle, are at right angles to each other.

Ex. 4. If the angle, between the internal bisector and a side, be equal to the angle, between the external bisector and the base, the perpendicular to the greater side, through the vertex, will bisect the segment of the base, cut off between the bisecting lines.

Proposition IV. Theorem.

The sides about the equal angles of triangles, which are equiangular to one another, are proportionals ; and those which are opposite to the equal angles, are homologous sides.

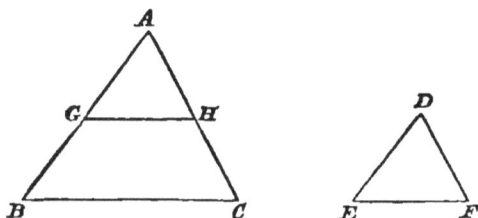

Let ABC, DEF be two $\triangle$s, having the $\angle$ s at A, B, C equal to the $\angle$ s at D, E, F respectively.

Then must the sides about the equal $\angle$ s be proportionals, those being homologous sides, which are opposite the equal $\angle$ s.

For suppose $\triangle$ DEF to be applied to $\triangle$ ABC,
so that D coincides with A and DE falls on AB ;
then $\because \angle BAC = \angle EDF$, $\therefore DF$ will fall on AC.

Let G and H be the points in AB and AC, or these produced, on which E and F fall.

Join GH. GH will be $\parallel$ to BC, $\because \angle AGH = \angle ABC$.   I. 28.

Then BA is to GA as CA is to HA,	VI. 2.
and $\therefore BA$ is to ED as CA is to FD,	V. 6.
whence BA is to AC as ED is to DF.	V. 15.

Similarly, by applying the $\triangle$ DEF, so that the $\angle$ s at F, E may coincide with those at C, B successively, we might show that

AC is to CB as DF is to FE, and that
CB is to BA as FE is to ED.

Q. E. D.

Ex. Divide a given angle into two parts, such that the perpendiculars from any point of the dividing line upon the two arms of the angle may be in a given ratio.

Proposition V. Theorem.

If the sides of two triangles, about each of their angles, be proportionals, the triangles must be equiangular to one another, and must have those angles equal, which are opposite to the homologous sides.

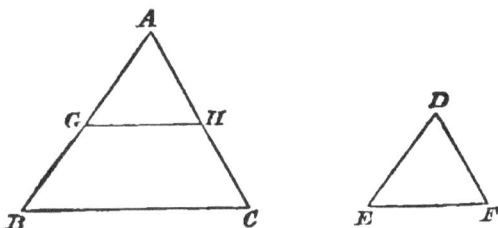

Let the △s *ABC, DEF* have their sides proportional,
so that *BA* is to *AC* as *ED* is to *DF*,
and *AC* is to *CB* as *DF* is to *FE*,
and *CB* is to *BA* as *FE* is to *ED*.

Then must △ *ABC be equiangular to* △ *EDF, those ∠s being equal, which are opposite to the homologous sides, that is,*
∠ *BAC* = ∠ *EDF*, and ∠ *ABC* = ∠ *DEF*, and ∠ *ACB* = ∠ *DFE*.

In *AB*, produced if necessary, make *AG* = *DE*,
and draw *GH* ∥ to *BC*, meeting *AC* in *H*.   I. 31.
Then △ *AGH* is equiangular to △ *ABC*,   I. 29.
and ∴ *BA* is to *AC* as *GA* is to *AH*.   VI. 4.
But *ED* is to *DF* as *BA* is to *AC*;   Hyp.
and ∴ *ED* is to *DF* as *GA* is to *AH*.   V. 5.
But *ED* = *GA*, and ∴ *DF* = *AH*.   V. 14.
So also it may be shown that *GH* = *EF*.
Then in △s *AGH, DEF*
∵ *GA* = *ED*, and *AH* = *DF*, and *HG* = *FE*,
∴ ∠ *GAH* = ∠ *EDF* ; ∠ *AGH* = ∠ *DEF* ; ∠ *AHG* = ∠ *DFE*.
          I. c.
But ∠ *GAH* = ∠ *BAC* ; ∠ *AGH* = ∠ *ABC* ; ∠ *AHG* = ∠ *ACB*.
∴ ∠ *BAC* = ∠ *EDF* ; ∠ *ABC* = ∠ *DEF*, and ∠ *ACB* = ∠ *DFE*.

Q. E. D.

Proposition VI. Theorem.

If two triangles have one angle of the one equal to one angle of the other, and the sides about the equal angles proportionals, the triangles must be equiangular to one another, and must have those angles equal, which are opposite to the homologous sides.

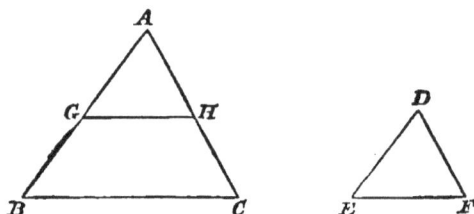

In the △s *ABC, DEF,* let ∠ *BAC=* ∠ *EDF,*
and let *BA* be to *AC* as *ED* to *DF.*

Then must △ *ABC be equiangular to* △ *DEF,*
and ∠ *ABC=* ∠ *DEF, and* ∠ *ACB=* ∠ *DFE.*

In *AB,* produced if necessary, make *AG=DE,*
and draw *GH* ∥ to *BC.*   I. 31.

Then △ *AGH* is equiangular to △ *ABC,*   I. 29.
and ∴ *GA* is to *AH* as *BA* is to *AC,*   VI. 4.
and ∴ *GA* is to *AH* as *ED* is to *DF.*   V. 5.

But *GA =ED,* by construction,
and ∴ *AH=DF.*   V. 14.

Then ∵ *GA =ED,* and *AH=DF* and ∠ *GAH=* ∠ *EDF;*
∴ ∠ *AGH=* ∠ *DEF,* and ∠ *AHG=* ∠ *DFE,*   I. 4.
and ∴ ∠ *ABC=* ∠ *DEF,* and ∠ *ACB=* ∠ *DFE,*

Q. E. D.

Ex. 1. If from *B, C,* the extremities of the base of a triangle *ABC,* be drawn *BD, CE,* perpendicular to the opposite sides, shew that the triangles *ADE, ABC* are equiangular.

Ex. 2. A variable chord *OP* is drawn through a fixed point *O* on the circumference of a circle, and *Q* is taken in it, so that the rectangle *OP, OQ* is constant, find the locus of *Q.*

Miscellaneous Exercises on Props. I. *to* VI.

1. If two triangles stand on the same base, and their vertices be joined by a straight line, the triangles are as the parts of this line intercepted between the vertices and the base.

2. If a circle be described on the radius of another circle as its diameter, and any straight line be drawn through the point of contact, cutting the two circles, the part, intercepted between the greater and lesser circles, shall be equal to the part within the lesser circle.

3. The side BC, of a triangle ABC, is bisected in D, and any straight line is drawn through D, meeting AB, AC, produced if necessary, in E, F, respectively, and the straight line through A, parallel to BC, in G. Prove that DE is to DF as GE is to GF.

4. If the angle A, of the triangle ABC, be bisected by AD, which cuts BC in D, and O be the middle point of BC, then OD bears the same ratio to OB that the difference of the sides bears to their sum.

5. The lines drawn from the base of a triangle perpendicular to the line bisecting the vertical angle, are in the same ratio as the sides of the triangle.

6. If D, E be points in the sides AB, AC respectively of the triangle ABC, such that the triangles DAC, EAB are equal, shew that the sides AB, AC are divided proportionally in D and E.

7. If two of the exterior angles, of a triangle ABC, be bisected by the lines COE, BOD, intersecting in O, and meeting the opposite sides in E and D, prove that OD is to OB as AD is to AB, and that OC is to OE as AC is to AE.

8. C, B, the angles at the base of an isosceles triangle, are joined to the middle points, E, F, of AB, AC, by lines intersecting in G. Shew that the area BCG is equal to the area $AEGF$.

9. If, through any point in the diagonal of a parallelogram, a straight line be drawn, meeting two opposite sides of the figure, the segments of this line will have the same ratio as those of the diagonal.

10. The sides AB, AC, of a triangle ABC, are produced to D and E, so that DE is parallel to BC, and the straight line DE is divided in F, so that DF is to FE as BD is to CE; shew that the locus of F is a straight line.

Proposition VII.   Theorem.

If two triangles have one angle of the one equal to one angle of the other, and the sides about a second angle in each proportionals ; then, if the third angles in each be both acute, both obtuse, or if one of them be a right angle, the triangles must be equiangular to one another, and must have those angles equal, about which the sides are proportionals.

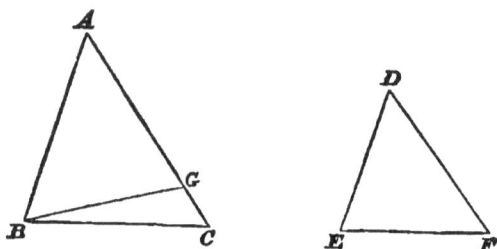

In the △s *ABC, DEF*, let ∠ *BAC=* ∠ *EDF*,

and let *AB* be to *BC* as *DE* is to *EF*,

and let ∠s *ACB, DFE* be both acute, both obtuse, or let one of them be a right angle.

Then must △s *ABC, DEF be equiangular to one another, having* ∠ *ABC=* ∠ *DEF, and* ∠ *ACB=* ∠ *DFE.*

For if ∠ *ABC* be not= ∠ *DEF*, let one of them, as ∠ *ABC*, be greater than the other, and make ∠ *ABG=* ∠ *DEF*,   I. 23.

and let *BG* meet *AC* in *G*.

Then ∵ ∠ *DAG=* ∠ *EDF*, and ∠ *ABG=* ∠ *DEF*,

∴ △ *ABG* is equiangular to △ *DEF*,   I. 32.

and ∴ *AB* is to *BG* as *DE* is to *EF*.   VI. 4.

But *AB* is to *BC* as *DE* is to *EF*,   Hyp.

∴ *AB* is to *BG* as *AB* is to *BC*,   V. 5.

and ∴ *BG=BC*,   V. 8.

and ∴ ∠ *BCG=* ∠ *BGC.*   I. A.

First, let ∠ *ACB* and ∠ *DFE* be both acute,

then ∠ *AGB* is acute, and ∴ ∠ *BGC* is obtuse ;   I. 13.

∴ ∠ *BCG* is obtuse, which is contrary to the hypothesis.

Next, let ∠ *ACB* and ∠ *DFE* be both obtuse,

then ∠ *AGB* is obtuse, and ∴ ∠ *BGC* is acute ;   I. 13.

∴ ∠ *BCG* is acute, which is contrary to the hypothesis.

Lastly, let one of the third ∠ s *ACB*, *DFE* be a right ∠ .

If ∠ *ACB* be a rt. ∠ ,

then ∠ *BGC* is also a rt. ∠ ;   I. A.

∴ ∠ s *BCG*, *BGC* together=two rt. ∠ s,

which is impossible.   I. 17.

Again, if ∠ *DFE* be a rt. ∠ ,

then ∠ *AGB* is a rt. ∠ , and ∴ ∠ *BGC* is a rt. ∠ .   I. 13.

Hence ∠ *BCG* is also a rt. ∠ ,   I. A.

and ∴ ∠ s *BCG*, *BGC* together=two rt. ∠ s,

which is impossible.   I. 17.

Hence ∠ *ABC* is not greater than ∠ *DEF*.

So also we might shew that ∠ *DEF* is not greater than ∠ *ABC*.

$$\therefore \angle ABC = \angle DEF,$$

and ∴ ∠ *ACB* = ∠ *DFE*.   I. 32.

Q. E. D.

N.B.—This Proposition is an extension of Proposition E of Book I. p. 42.

Note.—We have made a slight change in Euclid's arrangement of the four Propositions that follow, because Eucl. VI. 8 is closely connected with the proof of Eucl. VI. 13.

PROPOSITION VIII.   PROBLEM.   (Eucl. VI. 9.)

From a given straight line to cut off any submultiple.

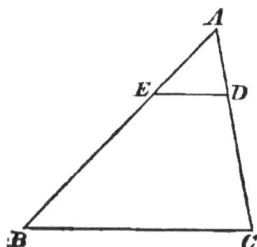

Let AB be the given st. line.

It is required to shew how to cut off any submultiple from AB.

From A draw AC making any angle with AB.

In AC take any pt. D, and make AC the same multiple of AD that AB is of the submultiple to be cut off from it.

Join BC, and draw $DE \parallel$ to BC.   I. 31.

Then ∵ ED is $\parallel$ to BC,

∴ CD is to DA as BE is to EA,   VI. 2.

and ∴ CA is to DA as BA is to EA.   V. 16.

∴ EA is the same submultiple of BA that DA is of CA
V. 19

Hence from AB the submultiple required is cut off.

Q. E. F.

Ex. 1. Cut off one-seventh of a given straight line.

Ex. 2. Cut off two-fifths of a given straight line.

Note.—This Proposition is a particular case of Proposition IX.

PROPOSITION IX.   PROBLEM.   (Eucl. VI. 10.)

To divide a given straight line similarly to a given straight line.

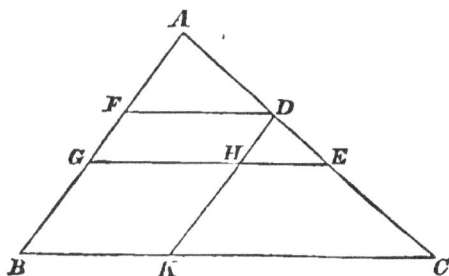

Let AB be the st. line given to be divided, and AC the divided st. line.

It is required to divide AB similarly to AC.

Let AC be divided in the pts. D, E.

Place AB, AC so as to contain any angle.

Join BC, and through D, E draw DF, $EG \parallel$ to BC.   I. 31.

Through D draw $DHK \parallel$ to AB.   I. 31.

Then $\because FH$ and GK are $\square$s,

$\therefore FG = DH$, and $GB = HK$.   I. 34.

And $\because HE$ is $\parallel$ to KC,

$\therefore KH$ is to HD as CE is to ED,   VI. 2.

that is, BG is to GF as CE is to ED.

Again, $\because FD$ is $\parallel$ to GE,

$\therefore GF$ is to FA as ED is to DA.   VI. 2.

Hence AB is divided similarly to AC.

Q. E. F.

Ex. 1. Produce a given straight line, so that the whole produced line shall be to the produced part in a given ratio.

Ex. 2. On a given base describe a triangle, with a given vertical angle and its sides in a given ratio.

PROPOSITION X.   PROBLEM.   (Eucl. VI. 11.)

To find a THIRD *proportional to two given straight lines.*

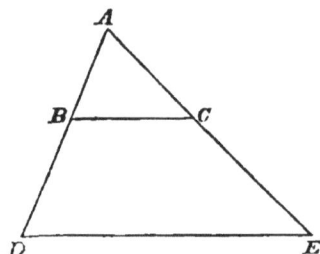

Let AB and AC be the given st. lines.

It is required to find a third proportional to AB, AC.

Place AB, AC so as to contain any angle.

Produce AB, AC to D and E, making $BD = AC$. I. 3.

Join BC, and through D draw $DE \parallel$ to BC.   I. 31.

Then $\because BC$ is $\parallel$ to DE,

$\therefore AB$ is to BD as AC is to CE,   VI. 2.

and $\therefore AB$ is to AC as AC is to CE.   V. 6.

Thus CE is a third proportional to AB and AC.

Q. E. F.

NOTE. This Proposition is a particular case of Proposition XI.

DEF. II. When three magnitudes are proportionals, the first is said to have to the third the duplicate ratio of that, which it has to the second.

Thus here AB has to CE the duplicate ratio of AB to AC.

DEF. III. When three magnitudes are proportionals, the first is said to have to the third the ratio compounded of the ratio, which the first has to the second, and of the ratio, which the second has to the third.

Thus here AB has to CE the ratio compounded of the ratios of AB to AC and AC to CE.

To find a FOURTH *proportional to three given straight lines.*

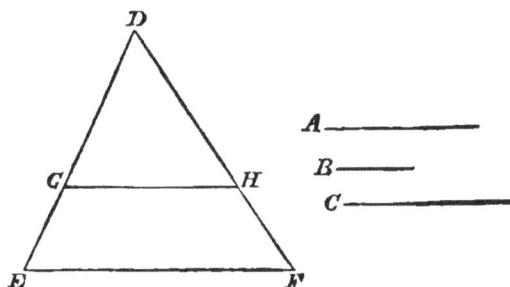

Let A, B, C be the three given st. lines.

It is required to find a fourth proportional to A, B, C.

Take DE, DF, two st. lines making an $\angle EDF$, and in these make $DG=A$, $GE=B$, and $DH=C$,   I. 3.

and through E draw $EF \parallel$ to GH.   I. 31.

Then, $\because GH$ is $\parallel$ to EF,

$\therefore DG$ is to GE as DH is to HF,   VI. 2.

and $\therefore A$ is to B as C is to HF.   V. 6.

Thus HF is a fourth proportional to A, B, C.

Q. E. F.

Ex. ABC is a triangle inscribed in a circle, and BD is drawn to meet the tangent to the circle at A in D, at an angle ABD equal to the angle ABC. Show that AC is a fourth proportional to the lines BD, DA, AB.

PROPOSITION XII.   THEOREM.   (Eucl. VI. 8.)

In a right-angled triangle, if a perpendicular be drawn from the right angle to the base, the triangles on each side of it are similar to the whole triangle and to one another.

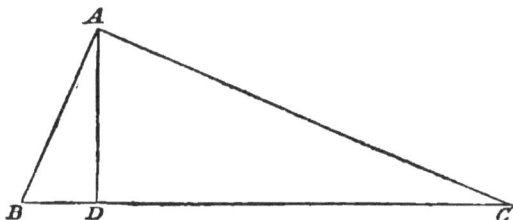

Let ABC be a right-angled $\triangle$, having $\angle BAC$ a rt. $\angle$, and from A let AD be drawn $\perp$ to BC.

Then must $\triangle s$ DBA, DAC *be similar to* $\triangle$ ABC, *and to each other.*

For ∵ rt. $\angle BDA$ =rt. $\angle BAC$, and $\angle ABD = \angle CBA$,

$$\therefore \angle DAB = \angle ACB. \qquad \text{I. 32.}$$

$\therefore \triangle DBA$ is equiangular, and $\therefore$ similar to $\triangle ABC$. VI. 4.

In the same way it may be shown

that $\triangle DAC$ is equiangular, and $\therefore$ similar to $\triangle ABC$.

Hence $\triangle DBA$ is similar to $\triangle DAC$.

Q. E. D.

COR. I. DA is a mean proportional between BD and DC,

For BD is to DA as DA is to DC.   VI. 4.

COR. II. BA is a mean proportional between BC and BD,

For BC is to BA as BA is to BD.   VI. 4.

COR. III. CA is a mean proportional between BC and CD,

For BC is to CA as CA is to CD.   VI. 4.

Q. E. D.

Ex. B is a fixed point in the circumference of a circle, whose centre is C; PA is a tangent at any point P, meeting CB produced in A, and PD is drawn perpendicularly to CB. Prove that the line bisecting the angle APD always passes through B.

To find a MEAN *proportional between two given straight lines.*

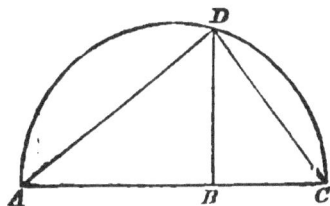

Let AB and BC be the two given st. lines.

It is required to find a mean proportional between AB *and* BC.

Place AB and BC so as to make one st. line AC,
  and on AC describe the semicircle ADC.

From B draw $BD \perp$ to AC, and join AD, CD.   I. 11.
    Then ∵ $\angle ADC$ is a rt. $\angle$,   III. 31.
     and DB is $\perp$ to AC,

∴ DB is a mean proportional between AB and BC.

               VI. 12, COR. 1.

            Q. E. F.

Ex. 1. Produce a given straight line, so that the given line may be a mean proportional between the whole line and the part produced.

Ex 2 Shew that either of the sides of an isosceles triangle is a mean proportional between the base and the half of the segment of the base, produced if necessary, which is cut off by a straight line, drawn from the vertex, at right angles to the equal side.

Ex. 3. Shew that the diameter of a circle is a mean proportional between the sides of an equilateral triangle and a hexagon, described about the circle.

Ex. 4. From a point A, outside a circle, a line is drawn, cutting the circle in B and C. Find a mean proportional between AB and AC.

DEF. IV. Two figures are said to have their sides about two of their angles *reciprocally proportional*, when, of the four terms of the proportion, the first antecedent and the second consequent are sides of one figure, and the second antecedent and first consequent are sides of the other figure.

Thus, in the diagram on the opposite page, the figures AB and BC have their sides about the angles at B reciprocally proportional, the order of the proportion being

$$DB \text{ is to } BE \text{ as } GB \text{ is to } BF.$$

Equal parallelograms, which have one angle of the one equal to one angle of the other, have their sides about the equal angles reciprocally proportional.

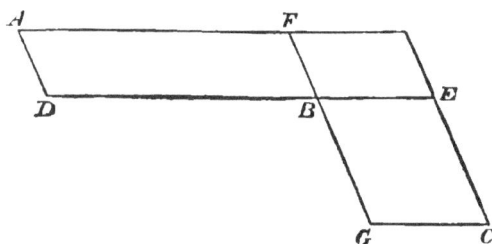

Let AB, BC be equal $\square$s, having $\angle FBD = \angle EBG$.

Then must DB be to BE as GB is to BF.

Place the $\square$s so that DB and BE are in the same st. line ; then must GB and BF also be in one st. line.　　　I. 14.

Complete the $\square$ *FE*.

Then $\because \square AB = \square BC$, and FE is another $\square$,

　$\therefore \square AB$ is to $\square FE$ as $\square BC$ is to $\square FE$.　　V. 6.

But as $\square AB$ is to $\square FE$ so is DB to BE,　VI. 1, Cor. I.

and as $\square BC$ is to $\square FE$ so is GB to BF.　VI. 1, Cor. I.

　$\therefore DB$ is to BE as GB is to BF.　　　V. 5.

And Conversely,

Parallelograms, which have one angle of the one equal to one angle of the other, and their sides about the equal angles reciprocally proportional, are equal to one another.

Let the sides about the equal $\angle$ s be reciprocally proportional, that is, let DB be to BE as GB is to BF.

Then must $\square AB = \square BC$.

For, the same construction being made,

　$\because DB$ is to BE as GB is to BF,

and that DB is to BE as $\square AB$ is to $\square FE$,　VI. 1, Cor. I.

and that GB is to BF as $\square BC$ is to $\square FE$,　VI. 1, Cor. I.

　$\therefore \square AB$ is to $\square FE$ as $\square BC$ is to $\square FE$.　V. 5.

　　and $\therefore \square AB = \square BC$.　　　V. 8.

Q. E. D.

Equal triangles, which have one angle of the one equal to one angle of the other, have their sides about the equal angles reciprocally proportional.

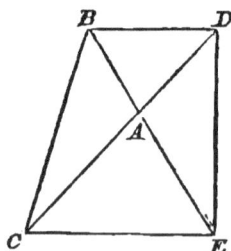

Let ABC, ADE be equal $\triangle$s, having $\angle BAC = \angle DAE$.

Then must CA be to AD as EA is to AB.

Place the $\triangle$s so that CA and AD are in the same st. line ; then must EA and AB also be in one st. line.   I. 14.

Join BD.

Then $\because \triangle ABC = \triangle ADE$, and ABD is another $\triangle$,

$\therefore \triangle ABC$ is to $\triangle ABD$ as $\triangle ADE$ is to $\triangle ABD$.   V. 6.

But as $\triangle ABC$ is to $\triangle ABD$ so is CA to AD,   VI. 1.

and as $\triangle ADE$ is to $\triangle ABD$ so is EA to AB.   VI. 1.

$\therefore CA$ is to AD as EA is to AB.   V. 5.

Ex. 1. Shew that, provided the sides of one of the triangles be made the extremes, it is indifferent, so far as the truth of the Proposition is concerned, in what order the sides of the other triangle are taken as the means of the four proportionals.

Ex. 2. ABb, AcC are two given straight lines, cut by two others BC, bc, so that the two triangles ABC, Abc may be equal ; shew that the lines BC, bc divide each other in reciprocal proportion.

And Conversely,

Triangles, which have one angle of the one equal to one angle of the other, and their sides about the equal angles reciprocally proportional, are equal to one another.

Let the sides about the equal $\angle$ s be reciprocally proportional, **that is, let CA be to AD as EA is to AB.**

Then must $\triangle ABC = \triangle ADE$.

For, the same construction being made,

$\because CA$ is to AD as EA is to AB,

and that CA is to AD as $\triangle ABC$ is to $\triangle ABD$,   VI. 1.

and that EA is to AB as $\triangle ADE$ is to $\triangle ABD$,   VI. 1.

$\therefore \triangle ABC$ is to $\triangle ABD$ as $\triangle ADE$ is to $\triangle ABD$.   V. 5.

and $\therefore \triangle ABC = \triangle ADE$.   V. 8.

Q. E. D.

Ex. 3. Through the extremities of the base BC, of a triangle ABC, draw two parallel lines, BE and CD, meeting AC and AB produced in E and D respectively, so that BCD may be equal in area to ABE.

Ex. 4. P is any point on the side AC, of the triangle ABC; CQ, drawn parallel to BP, meets AB produced in Q; AN, AM are mean proportionals between AB, AQ, and AC, AP, respectively. Shew that the triangle ANM is equal to the triangle ABC.

PROPOSITION XVI.   THEOREM.

If four straight lines be proportionals, the rectangle contained by the extremes is equal to the rectangle contained by the means.

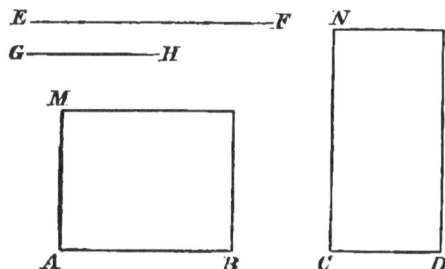

Let the four st. lines AB, CD, EF, GH be proportionals, so that AB is to CD as EF is to GH.

Then must rect. AB, GH = rect. CD, EF.

Draw $AM \perp$ to AB, and $CN \perp$ to CD ;   I. 11.

and make $AM = GH$, and $CN = EF$;

and complete the $\square$s BM, DN.   I. 31.

Then ∵ AB is to CD as EF is to GH,

and that $EF = CN$, and $GH = AM$,

∴ AB is to CD as CN is to AM.   V. 6.

Thus the sides about the equal ∠ s of the equiangular $\square$s BM, DN are reciprocally proportional,

and ∴ $\square$ BM = $\square$ DN ;   VI. 14.

that is, rect. AB, AM = rect. CD, CN.

∴ rect. AB, GH = rect. CD, EF.

Ex. 1. If E be the middle point of a semicircular arc AEB, and EDC be any chord, cutting the diameter in D, and the circle in C, prove that the square on CE is equal to twice the quadrilateral $AEBC$.

And Conversely,

If the rectangle contained by the extremes be equal to the rectangle contained by the means, the four straight lines are proportionals.

Let rect. AB, GH = rect. CD, EF.

Then must AB be to CD as EF is to GH.

For, the same construction being made,

$\because$ rect. AB, GH = rect. CD, EF,

$\therefore$ rect. AB, AM = rect. CD, CN,

that is, $\square$ BM = $\square$ DN.

and these $\square$s are equiangular to one another,

and $\therefore$ the sides about the equal $\angle$ s are reciprocally proportional,   VI. 14.

and $\therefore$ AB is to CD as CN is to AM,

and $\therefore$ AB is to CD as EF is to GH.   V. 6.

Q. E. D.

Ex. 2. If, from an angle of a triangle, two straight lines be drawn, one to the side subtending that angle, and the other cutting from the circumscribing circle a segment, capable of containing an angle, equal to the angle, contained by the first drawn line and the side, which it meets ; the rectangle, contained by the sides of the triangle, shall be equal to the rectangle, contained by the lines thus drawn.

PROPOSITION XVII.   THEOREM.

If three straight lines be proportionals, the rectangle contained by the extremes is equal to the square on the mean.

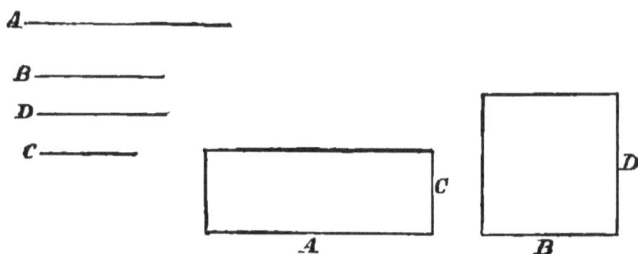

Let the three st. lines *A*, *B*, *C* be proportionals, and let *A* be to *B* as *B* is to *C*.

Then must rect. A, C=sq. on B.

Take *D*=*B*.

Then ∵ *A* is to *B* as *B* is to *C*,

∴ *A* is to *B* as *D* is to *C*,   V. 6.

and ∴ rect. *A*, *C*=rect. *B*, *D*,   VI. 16.

that is, rect. *A*, *C*=sq. on *B*.

And Conversely,

If the rectangle contained by the extremes be equal to the square on the mean, the three straight lines are proportionals.

Let *A*, *B*, *C* be three straight lines such that

rect. *A*, *C*=sq. on *B*.

Then must A be to B as B is to C.

For, the same construction being made,

∵ rect. *A*, *C*=sq. on *B*,

and *B*=*D*,

∴ rect. *A*, *C*=rect. *B*, *D* ,

and ∴ *A* is to *B* as *D* is to *C*,   VI. 16.

that is, *A* is to *B* as *B* is to *C*.   V. 6.

Q. E. D.

Upon a given straight line to describe a rectilinear figure similar and similarly situated to a given rectilinear figure.

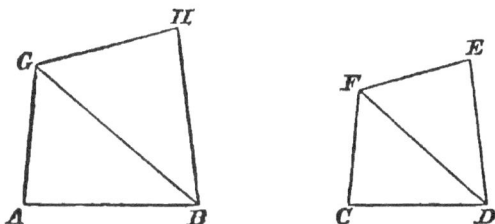

Let AB be the given st. line, and $CDEF$ the given rectil. fig. of *four* sides.

It is required to describe on AB a fig. similar and similarly situated to $CDEF$.

Join DF, and at A and B, make $\angle BAG = \angle DCF$, and $\angle ABG = \angle CDF$;

then $\triangle BAG$ is equiangular to $\triangle DCF$.

At G and B, make $\angle BGH = \angle DFE$, and $\angle GBH = \angle FDE$;

then $\triangle GHB$ is equiangular to $\triangle FED$.

Then $\because \angle AGB = \angle CFD$, and $\angle BGH = \angle DFE$,

$$\therefore \angle AGH = \angle CFE. \qquad \text{Ax. 2.}$$

So also $\angle ABH = \angle CDE$.

And we know that $\angle BAG = \angle DCF$,

and that $\angle GHB = \angle FED$,

$\therefore$ rectil. fig. $ABHG$ is equiangular to fig. $CDEF$.

Also, $\because \triangle BAG$ is equiangular to $\triangle DCF$,

$$\therefore BA \text{ is to } AG \text{ as } DC \text{ is to } CF; \qquad \text{VI. 4.}$$

and $\because \triangle BGH$ is equiangular to $\triangle DFE$,

$$\therefore GB \text{ is to } GH \text{ as } FD \text{ is to } FE. \qquad \text{VI. 4.}$$

Also, AG is to GB as CF is to FD.

$$\therefore AG \text{ is to } GH \text{ as } CF \text{ is to } FE. \qquad \text{V. 21.}$$

Similarly, it may shown that

GH is to HB as FE is to ED,

and that HB is to BA as ED is to DC.

$\therefore$ the rectil. figs. $ABHG$ and $CDEF$ are similar.

NEXT. Let it be required to describe on AB a fig., similar and similarly situated to the rectil. fig. $CDKEF$.

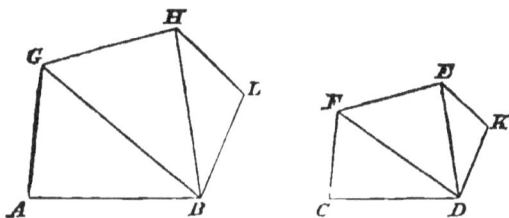

Join DE, and on AB describe the fig. $ABHG$, similar and similarly situated to the quadrilateral $CDEF$.

At B and H make $\angle HBL = \angle EDK$, and $\angle BHL = \angle DEK$; then $\triangle HLB$ is equiangular to $\triangle EKD$.

Then $\because$ the figs. $ABHG$, $CDEF$ are similar,

$$\therefore \angle GHB = \angle FED ;$$

and we have made $\angle BHL = \angle DEK$;

$$\therefore \text{ whole } \angle GHL = \text{whole } \angle FEK. \qquad \text{Ax. 2.}$$

For the same reason, $\angle ABL = \angle CDK$.

Thus the fig. $AGHLB$ is equiangular to fig. $CFEKD$.

Again, $\because$ the figs. $AGHB$, $CFED$ are similar,

$$\therefore GH \text{ is to } HB \text{ as } FE \text{ is to } ED :$$

also we know that HB is to HL as ED is to EK, VI. 4.

$$\therefore GH \text{ is to } HL \text{ as } FE \text{ is to } EK. \qquad \text{V. 21.}$$

For the same reason, AB is to BL as CD is to DK.

And BL is to LH as DK is to KD; VI. 4.

$$\therefore \text{ the five-sided figs. } AGHLB, CFEKD \text{ are similar.}$$

In the same way a fig. of six or more sides may be described on a given line, similar to a given fig.

Q. E. F.

Similar triangles are to one another in the duplicate ratio of their homologous sides.

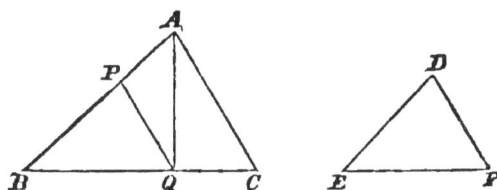

Let ABC, DEF be similar $\triangle$s,
having $\angle$ s at A, B, $C=\angle$ s at D, E, F respectively,
so that BC and EF are homologous sides.

Then must $\triangle\,ABC$ have to $\triangle\,DEF$ the duplicate ratio of that which BC has to EF.

Suppose $\triangle\,DEF$ to be applied to $\triangle\,ABC$, so that E lies on B, ED on BA, and $\therefore EF$ on BC.

Let P and Q be the pts. in BA, BC on which D and F fall.

Join AQ.

Then $\triangle\,ABC$ is to $\triangle\,ABQ$ as BC is to BQ,	VI. 1.
and $\triangle\,ABQ$ is to $\triangle\,PBQ$ as AB is to BP.	VI. 1.
But AB is to BP as BC is to BQ,	VI. 4.
$\therefore \triangle\,ABQ$ is to $\triangle\,PBQ$ as BC is to BQ.	V. 5.
Hence $\triangle\,ABC$ is to $\triangle\,ABQ$ as $\triangle\,ABQ$ is to $\triangle\,PBQ$.	V. 5.

$\therefore \triangle\,ABC$ has to $\triangle\,PBQ$ the duplicate ratio
of $\triangle\,ABC$ to $\triangle\,ABQ$;   VI. Def. 2.

$\therefore \triangle\,ABC$ has to $\triangle\,PBQ$ the duplicate ratio
of BC to BQ.   V. 5.

that is, $\triangle\,ABC$ has to $\triangle\,DEF$ the duplicate ratio
of BC to EF.

Q. E. D.

COR. If MN be a third proportional to BC and EF,
BC has to MN the duplicate ratio of BC to EF,   VI. Def. 2.
and $\therefore BC$ is to MN as $\triangle\,ABC$ is to $\triangle\,DEF$.

Miscellaneous Exercises chiefly on Proposition XIX.

Ex. 1. Prove this Proposition without drawing any line inside either of the triangles.

Ex. 2. In the figure, if BC be equal to FD, shew that the triangles will be in the ratio of AC to EF.

Ex. 3. Cut off the third part of a triangle by a straight line parallel to one of its sides.

Ex. 4. AB, AC are bisected in D and E. Prove that the quadrilateral $DBCE$ is equal to three times the triangle ADE.

Ex. 5. If a regular hexagon, a square, and an equilateral triangle be inscribed in the same circle, prove that the squares described on their sides are proportional to the numbers 1, 2, 3.

Ex. 6. A straight line drawn parallel to the diagonal BD of a parallelogram $ABCD$ meets AB, BC, CD, DA, in E, F, G, H. Prove that the triangles AFG, CEH are equal.

Ex. 7. If two triangles have an angle equal, and be to each other in the duplicate ratio of adjacent sides, they are similar.

Ex. 8. If two triangles have a common angle, shew that the areas of the triangles are proportional to the rectangles contained by the sides of the triangles about the common angle.

Ex. 9. From the extremities A, B, of the diameter of a circle, perpendiculars AY, BZ, are let fall on the tangent at any point C. Prove that the areas of the triangles ACY, BCZ are together equal to that of the triangle ACB.

Ex. 10. If to the circle, circumscribing the triangle ABC, a tangent at C be drawn, cutting AB produced in D, shew that AD is to DB in the duplicate ratio of AC to CB.

Ex. 11. Construct a triangle which shall be to a given triangle in a given ratio.

PROPOSITION XX. THEOREM. (Eucl. VI. 21.)

Rectilinear figures, which are similar to the same rectilinear figure, are also similar to each other.

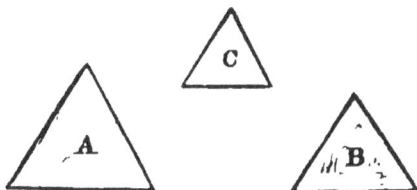

Let each of the rectilinear figures *A* and *B* be similar to the rectilinear figure *C*.

Then must the figure A be similar to the figure B.

For ∵ *A* is similar to *C*,

∴ *A* is equiangular to *C*,

and *A* and *C* have their sides about the equal ∠s proportionals.   VI. Def. 1.

Again, ∵ *B* is similar to *C*,

∴ *B* is equiangular to *C*,

and *B* and *C* have their sides about the equal ∠s proportionals.   VI. Def. 1.

Hence *A* and *B* are each equiangular to *C*, and have the sides about the equal ∠s of each of them and of *C* proportionals.

∴ *A* is equiangular to *B*,   Ax. 1.

and *A* and *B* have their sides about the equal ∠s proportionals.   V. 5.

∴ the figure *A* is similar to the figure *B*. VI. Def. 1.

Q. E. D.

PROPOSITION XXI. THEOREM. (Eucl. VI. 20.)

Similar polygons may be divided into the same number of similar triangles, having the same ratio to one another, which the polygons have; and the polygons are to one another in the duplicate ratio of their homologous sides.

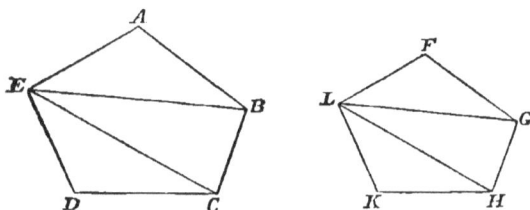

Let *ABCDE, FGHKL* be similar polygons, and let *AB* be the side homologous to *FG.*

I. *The polygons may be divided into the same number of similar △ s.*

II. *These △ s have each to each the same ratio which the polygons have.*

III. *The polygon ABCDE has to the polygon FGHKL the duplicate ratio of that which the side AB has to the side FG.*

Join *BE, EC, GL, LH* : then

I. ∵ the polygon *ABCDE* is similar to the polygon *FGHKL,*

∴ ∠ *BAE = ∠ GFL,*

and *BA* is to *AE* as *GF* is to *FL.*

∴ △ *ABE* is similar to △ *FGL.*   VI. 6 and 4

and ∴ ∠ *ABE = ∠ FGL.*   VI. Def. 1.

Again, ∵ the polygons are similar,

∴ ∠ *ABC = ∠ FGH,*   VI. Def. 1.

and ∴ ∠ *EBC = ∠ LGH* ;   A x. 3.

and ∵ the △ s *ABE, FGL* are similar,

∴ *EB* is to *AB* as *LG* is to *FG* ;   VI. Def. 1.

also, ∵ the polygons are similar,

∴ *AB* is to *BC* as *FG* is to *GH* ;   VI. Def. 1.

and ∴ *EB* is to *BC* as *LG* is to *GH,*   V. 21.

and ∴ since ∠ *EBC = ∠ LGH,*

the △ *EBC* is similar to △ *LGH.*   VI. 6 and 4

For the same reason the △ *ECD* is similar to △ *LHK.*

Thus the polygons are divided into the same number of similar △ s.

II. ∵ △ *ABE* is similar to △ *FGL*,

∴ △ *ABE* has to △ *FGL* the duplicate ratio of
BE to *GL*. VI. 19.

So also, △ *EBC* has to △ *LGH* the duplicate ratio of
BE to *GL*. VI. 19.

∴ △ *ABE* is to △ *FGL* as △ *EBC* is to △ *LGH*. V. 5.

Again, ∵ △ *EBC* is similar to △ *LGH*,

∴ △ *EBC* has to △ *LGH* the duplicate ratio of
EC to *LH*. VI. 19.

So also, △ *ECD* has to △ *LHK* the duplicate ratio of
EC to *LH*. VI. 19.

∴ △ *EBC* is to △ *LGH* as △ *ECD* is to △ *LHK*. V. 5.

But △ *EBC* is to △ *LGH* as △ *ABE* is to △ *FGL*.

∴ as △ *ABE* is to △ *FGL* so is △ *EBC* to △ *LGH*,

and △ *ECD* to △ *LHK*.

Now as one of the antecedents is to one of the consequents
so are all the antecedents together to all the consequents
together, V. 10.

and ∴ △ *ABE* is to △ *FGL* as polygon *ABCDE* is to polygon
FGHKL.

III. Since △ *ABE* has to △ *FGL* the duplicate ratio of
AB to *FG*, VI. 19.

∴ polygon *ABCDE* has to polygon *FGHKL* the duplicate
ratio of *AB* to *FG*. V. 5.

Q. E. D.

COR. I. In like manner it may be proved, that similar
figures of *four* or *any number* of sides, are to one another in
the duplicate ratio of their homologous sides : and it has been
already proved for *triangles*, VI. 19. Therefore, universally,
similar rectilinear figures are to one another in the duplicate
ratio of their homologous sides.

Cor. II. If MN be a third proportional to AB and FG, AB has to MN the duplicate ratio of AB to FG,   VI. Def. 2. and ∴ AB is to MN as the figure on AB to the similar and similarly described figure on FG ; that being true in the case of quadrilaterals and polygons, which has been already proved for triangles.   VI. 19 Cor.

PROPOSITION XXII.   THEOREM.   (Eucl. VI. 31.)

In right-angled triangles, the rectilinear figure, described upon he side opposite to the right angle, is equal to the similar and similarly described figures upon the sides containing the right angle.

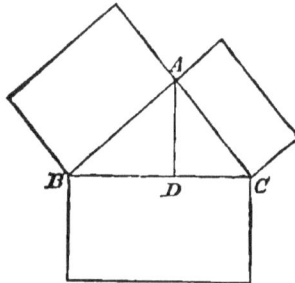

Let ABC be a right-angled △, having the right ∠ BAC. *Then must the rectilinear figure, described on BC, be equal to the similar and similarly described figures on BA, AC.*

Draw $AD \perp$ to BC.

Then △ ABC is similar to △ DBA,   VI. 12.

and ∴ BC is to BA as BA is to BD,   VI. 4.

and ∴ as BC is to BD so is the figure described on BC to the similar and similarly described figure on BA, VI. 21, Cor. 2

and ∴ as BD is to BC so is figure on BA to figure on BC.
                                                            V. 12

For the same reason

as DC is to BC so is figure on AC to figure on BC.

Hence as BD, DC together are to BC so are figures on BA. AC together to figure on BC.   V. 22.

But BD, DC together are equal to BC, and

∴ figures on BA, AC together = figure on BC. V. 18.

                                               Q. E. D.

NOTE.—The Proposition which follows is not given by Euclid, but is necessary to the proof of Prop. XXIV.

PROPOSITION XXIII. THEOREM.

If two rectilinear figures be equal and also similar, their homologous sides must be equal, each to each.

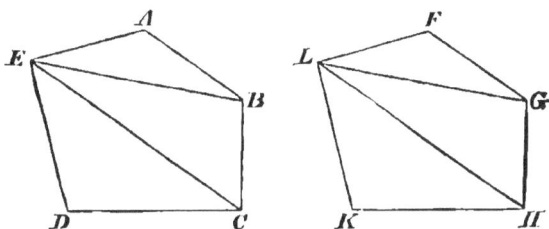

Let the rectil. figs. $ABCDE$, $FGHKL$ be equal and similar, and let DC and KH be homologous sides of the figures.

$$\text{Then must } DC = KH.$$

For, if not, let DC be greater than KH.

Then $\because$ DC is to DE as KH is to KL,

$\therefore DE$ is greater than KL.　　　　　　　　　V. 14.

Hence if $\triangle KLH$ be applied to $\triangle DEC$, so that KH falls on DC and KL on DE (for $\angle HKL = \angle CDE$), HL will fall entirely within $\triangle DEC$,

$\therefore \triangle KLH$ is less than $\triangle DEC$.

But $\because \triangle DEC$ is to $\triangle KLH$ as figure $ABCDE$ is to figure $FGHKL$,　　　　　　　　　　VI. 21.

and figure $ABCDE$ = figure $FGHKL$

$\therefore \triangle DEC = \triangle KLH$,　　　　　　　　　　V. 18.

or the greater = the less, which is impossible.

$\therefore DC$ is not greater than KH.

Similarly it may be shown that DC is not less than KH.

$\therefore DC = KH$.

Q. E. D.

PROPOSITION XXIV.   (Eucl. VI. 22.)

*If four straight lines be proportionals, the similar recti-
linear figures similarly described upon them must also be proportionals.*

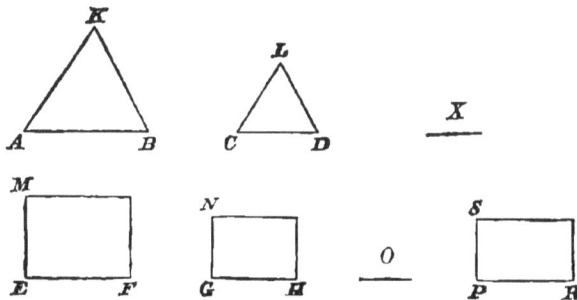

Let the four straight lines *AB*, *CD*, *EF*, *GH* be proportionals, that is, *AB* to *CD* as *EF* is to *GH* ;

and upon *AB*, *CD* let the similar rectilinear figures *KAB*,
LCD be similarly described ; and upon *EF*, *GH* the similar
rectilinear figures *MF*, *NH* in like manner.

Then must KAB be to LCD as MF is to NH.

To *AB*, *CD* take a third proportional *X* and

to *EF*, *GH* take a third proportional *O*.   VI. 10.

Then ∵ *AB* is to *CD* as *EF* is to *GH*,

∴ *CD* is to *X* as *GH* is to *O*,   V. 5.

and ∴ *AB* is to *X* as *EF* is to *O*.   V. 21.

But as *AB* is to *X* so is *KAB* to *LCD*, VI. 21, Cor. 2.

and as *EF* is to *O* so is *MF* to *NH*. VI. 21, Cor. 2.

∴ *KAB* is to *LCD* as *MF* is to *NH*.   V. 5.

And Conversely,

If the similar figures, similarly described on four straight lines, be proportionals, those straight lines must be proportionals.

The same construction being made,

let *KAB* be to *LCD* as *MF* is to *NH*,

then must AB be to CD as EF is to GH.

Make as *AB* to *CD* so *EF* to *PR*, VI. 11.

and on *PR* describe the rectilinear figure *SR*, similar and similarly situated to either of the figures *MF*, *NH*. VI. 18.

Then, by the first part of the proposition,

KAB is to *LCD* as *MF* is to *SR*.

But *KAB* is to *LCD* as *MF* is to *NH*. Hyp.

∴ *SR* = *NH*, V. 8.

Also, *SR* and *NH* are similar and similarly situated,

and ∴ *PR* = *GH*. VI. 23.

Now *AB* is to *CD* as *EF* is to *PR*,

and ∴ *AB* is to *CD* as *EF* is to *GH*. V. 6.

Q. E. D.

PROPOSITION XXV. THEOREM. (Eucl. VI. 33.)

In equal circles, angles, whether at the centres or the circumferences, have to one another the same ratio as the arcs which subtend them; and so also have the sectors.

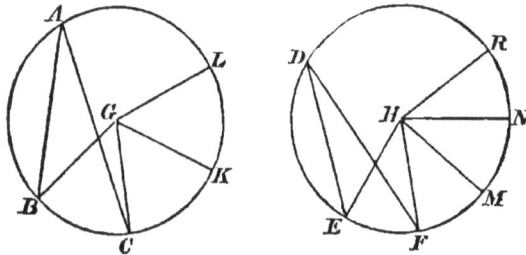

In the equal ⊙s *ABC, DEF* let the ∠s *BGC, EHF* at the centres, and the ∠s *BAC, EDF* at the circumferences, be subtended by the arcs *BC, EF*.

Then I. ∠ *BGC must be to* ∠ *EHF as arc BC is to arc EF.*

Take any number of arcs *CK, KL*, each = *BC*,
and any number of arcs *FM, MN, NR* each = *EF*.
Then ∵ arcs *BC, CK, KL* are all equal,
∴ ∠s *BGC, CGK, KGL* are all equal.   III. 27.
∴ ∠ *BGL* is the same multiple of ∠ *BGC* that arc *BL* is of arc *BC*.
So also, ∠ *EHR* is the same multiple of ∠ *EHF* that arc *ER* is of arc *EF*.
    And ∠ *BGL* is equal to, greater than, or less than ∠ *EHR*,
    according as arc *BL* is equal to, greater than, or less than arc *ER*.   III. 27.
Now ∠ *BGL* and arc *BL* are equimultiples of ∠ *BGC* and arc *BC*,
and ∠ *EHR* and arc *ER* are equimultiples of ∠ *EHF* and arc *EF*.
∴ ∠ *BGC* is to ∠ *EHF* as arc *BC* is to arc *EF*.   V. Def. 5.

II. ∠ *BAC* must be to ∠ *EDF* as arc *BC* is to arc *EF*.

For ∵ ∠ *BGC*=twice ∠ *BAC*, and ∠ *EHF*=twice ∠ *EDF*,

III. 20.

∴ ∠ *BAC* is to ∠ *EDF* as ∠ *BGC* is to ∠ *EHF*,   V. 11.

and ∴ ∠ *BAC* is to ∠ *EDF* as arc *BC* is to arc *EF*. V. 5.

III. *Sector BGC must be to sector EHF as arc BC is to arc EF.*

For sectors *BGC*, *CGK*, *KGL* are all equal,   III. 26, Cor.

and sectors *EHF*, *FHM*, *MHN*, *NHR*, are all equal,

III. 26, Cor.

∴ sector *BGL* is the same multiple of sector *BGC* that ιrc *BL* is of arc *BC*,

and sector *EHR* is the same multiple of sector *EHF* that arc *ER* is of arc *EF* ;

also, sector *BGL* is equal to, greater than or less than sector *EHR*, according as

arc *BL* is equal to, greater than, or less than arc *ER*, III. 26.

and ∴ sector *BGC* is to sector *EHF* as arc *BC* is to arc *EF*.

Q. E. D.

COR. In the *same* circle, angles, whether at the centres or the circumferences, have the same ratio as the arcs which subtend them ; and so also have the sectors.

Proposition B. Theorem.

If an angle of a triangle be bisected by a straight line, which likewise cuts the base ; the rectangle, contained by the sides of the triangle, is equal to the rectangle, contained by the segments of the base, together with the square on the line bisecting the angle.

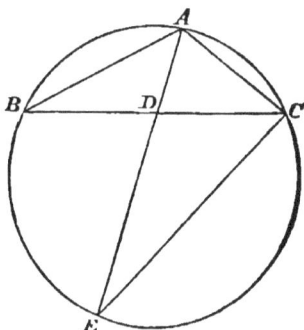

Let $\angle BAC$ of the $\triangle ABC$ be bisected by the st. line AD.

Then rect. BA, AC=rect. BD, DC together with sq. on AD.

Describe the $\odot$ ABC about the $\triangle$,   III. B. p. 135.

   produce AD to meet the $\bigcirc$ce in E, and join EC.

Then $\because \angle BAD = \angle CAE$,   Hyp.

and $\angle ABD = \angle AEC$, in the same segment,   III. 21.

$\therefore \triangle ABD$ is equiangular to $\triangle AEC$.   I. 32.

$\therefore BA$ is to AD as EA is to AC.   VI 4.

$\therefore$ rect. BA, AC=rect. EA, AD,   VI. 16.

                    =rect. ED, DA together with sq. on AD.
                                        II. 3.

                    =rect. BD, DC together with sq. on AD.
                                        III. 35.

                                        Q. E. D.

PROPOSITION C. THEOREM.

If from any angle of a triangle a straight line be drawn perpendicular to the base, the rectangle, contained by the sides of the triangle, is equal to the rectangle, contained by the perpendicular and the diameter of the circle described about the triangle.

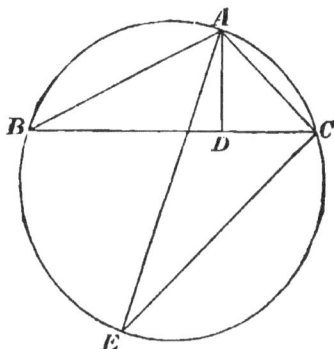

Let ABC be a $\triangle$, and AD the $\perp$ from A to BC.

Describe the $\odot$ ABC about the $\triangle$ ABC,	**III. s.**
draw the diameter AE, and join EC.	
Then must rect. BA, $AC =$ rect. EA, AD.	
For $\because$ rt. $\angle$ $BDA = \angle$ ECA, in a semicircle,	III. 31.
and $\angle$ $ABD = \angle$ AEC, in the same segment,	III. 21.
$\therefore$ $\triangle$ ABD is equiangular to the $\triangle$ AEC.	I. 32.
$\therefore$ BA is to AD as EA is to AC,	VI. 4.
and $\therefore$ rect. BA, $AC=$rect. EA, AD.	VI. 16.

Q. E. D.

Ex. 1. Shew that the rectangle contained by the two sides can never be less than twice the triangle.

Ex. 2. ABC is a triangle, and AM the perpendicular upon BC, and P any point in BC; if O, O' be the centres of the circles described about ABP, ACP, the rectangle AP, BC is double of the rectangle of AM, OO'.

Ex. 3. A bisector of an angle of a triangle is produced to meet the circumscribed circle. Prove that the rectangle, contained by this whole line and the part of it within the triangle, is equal to the rectangle contained by the two sides.

PROPOSITION D. THEOREM.

The rectangle, contained by the diagonals of a quadrilateral inscribed in a circle, is equal to the sum of the rectangles, contained by its opposite sides.

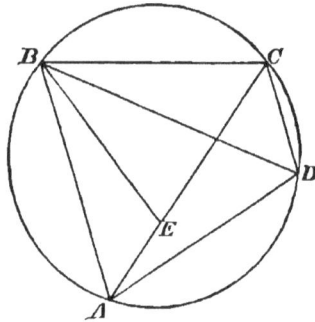

Let $ABCD$ be any quadrilateral inscribed in a ⊙.
Join AC, BD.

Then rect. AC, BD=rect. AB, CD together with rect. AD, BC.

Make $\angle ABE = \angle DBC$;   I. 23.

and add to each the $\angle EBD$.

Then $\angle ABD = \angle CBE$;

and $\angle BDA = \angle BCE$ in the same segment ;   III. 21.

∴ △ ABD is equiangular to △ BCE,   I. 32.

∴ AD is to BD as CE is to BC,   VI. 4.

and ∴ rect. AD, BC=rect. BD, CE.   VI. 16.

Again, ∵ $\angle ABE = \angle DBC$, by construction,

and $\angle BAE = \angle BDC$, in the same segment,   III. 21.

∴ △ ABE is equiangular to △ BCD.   I. 32.

∴ AB is to AE as BD is to CD,   VI. 4,

and ∵ rect. AB, CD = rect. BD, AE.   VI. 16.

Hence rect. AB, CD together with rect. AD, BC
=rect. BD, AE together with rect. BD, CE.
=rect. AC, BD.   II. 1.

Q. E. D.

Ex. If the diagonals cut one another at an angle equal to one third of a right angle, the rectangles contained by the opposite sides are together equal to four times the quadrilateral figure.

Equiangular parallelograms have to one another the ratio, which is compounded of the ratios of their sides.

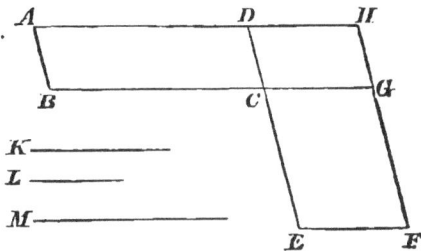

Let AC and CF be equiangular ▱s, having ∠ $BCD =$ ∠ ECG.

Then must ▱ AC *have to* ▱ CF *the ratio compounded of the ratios of their sides.*

Let BC and CG be placed in a straight line.

Then DC and CE are also in a straight line.   I. 14.

Complete the ▱ DG, and taking any st. line K,

    make as BC is to CG so K to L VI. 11.

    and make as DC is to CE so L to M.   VI. 11.

Then ∵ K has to M the ratio compounded of the ratios of K to L and L to M,

   ∴ K has to M the ratio compounded of the ratios of the sides.   VI. Def. 3, p. 260.

   Now BC is to CG as ▱ AC is to ▱ CH,   VI. 1.

   and DC is to CE as ▱ CH is to ▱ CF,   VI. 1.

   ∴ K is to L as ▱ AC is to ▱ CH,   V. 5.

   and L is to M as ▱ CH is to ▱ CF,   V. 5.

   Hence K is to M as ▱ AC is to ▱ CF ;   V. 21.

and ∴ ▱ AC has to ▱ CF the ratio compounded of the ratios of their sides.

         Q. E. D.

PROPOSITION XXVII. THEOREM. (Eucl. VI. 24).

Parallelograms about the diameter of any parallelogram are similar to the whole parallelogram and to one another.

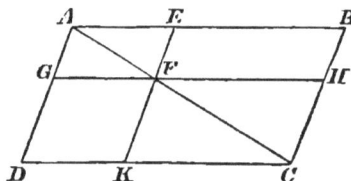

Let $ABCD$ be a $\square$, of which the diameter is AC; and $AEFG$, $FHCK$ the $\square$s about the diameter.

Then must these $\square$s be similar to ABCD and to each other.

For ∵ GF is ∥ to DC, ∴ ∠ AGF = ∠ ADC,   I. 29.

and ∵ EF is ∥ to BC, ∴ ∠ AEF = ∠ ABC;   I. 29.

and each of the ∠ s EFG, BCD = opposite ∠ BAD,   I. 34.

and ∴ ∠ EFG = ∠ BCD.   Ax. 1.

Thus the $\square$s $AEFG$, $ABCD$ are equiangular to one another.

Again, ∵ EF is ∥ to BC,

∴ AB is to BC as AE is to EF;   VI. 4.

and since the opposite sides of the $\square$s are equal,

∴ AB is to AD as AE is to AG,   V. 6.

and DC is to CB as GF is to FE,   V. 6.

and CD is to DA as FC is to GA.   V. 6.

Thus the sides of the $\square$s $AEFG$, $ABCD$ about their equal angles are proportional.

∴ $\square$ $AEFG$ is similar to $\square$ $ABCD$.

Similarly, $\square$ $FHCK$ is similar to $\square$ $ABCD$;

and ∴ $\square$ $AEFG$ is similar to $\square$ $FHCK$.   VI. 20.

Q. E. D.

Ex. Show that each of the complements of the parallelogram is a mean proportional between the parallelograms about the diameter.

PROPOSITION XXVIII. THEOREM. (Eucl. vi. 26.)

If two similar parallelograms have a common angle, and be similarly situated, they are about the same diameter.

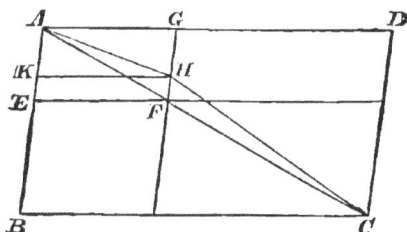

Let the ▱s *ABCD*, *AEFG* be similar and similarly situated, and have ∠ *DAB* common.

Then must ABCD and AEFG be about the same diameter.

For, if not, let *ABCD* have its diameter, *AHC*, not in the same st. line with *AF*, the diameter of *AEFG*.

Let *GF* meet *AHC* in *H*, and draw *HK* ‖ to *AD*.   I. 31.

Then ▱s *ABCD*, *AKHG*, about the same diameter, are similar.   VI. 27.

and ∴ *DA* is to *AB* as *GA* is to *AK*.   VI. Def. 1.

But ∵ *ABCD*, *AEFG* are similar ▱s,

∴ *DA* is to *AB* as *GA* is to *AE*.

Hence *GA* is to *AK* as *GA* is to *AE*,   V. 5.

and ∴ *AK* = *AE*,   V. 8.

the less = the greater, which is impossible.

∴ *ABCD* and *AKHG* are not about the same diameter, and ∴ *ABCD* and *AEFG* must have their diameters in the same st. line, that is, they are about the same diameter.

Q. E. D.

PROPOSITION XXIX.   PROBLEM.   (Eucl. VI. 25.)

To describe a rectilinear figure which shall be similar to one, and equal to another, given rectilinear figure.

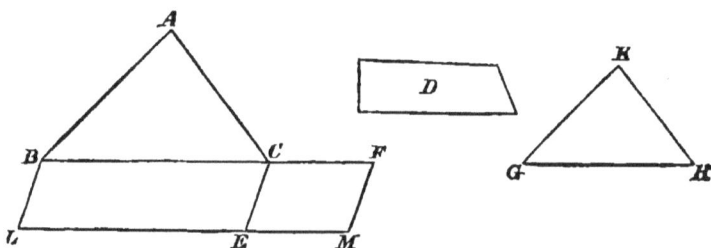

Let *ABC* and *D* be two given rectilinear figures.

It is required to describe a figure similar to ABC and equal to D.

On *BC* describe the ☐ *BLEC* equal to *ABC*, and   I. 45, Cor.
on *CE* describe the ☐ *CEFM* equal to *D*,   I. 45, Cor.
                and having $\angle FCE = \angle CBL$.
    Then *BC* and *CF* are in a straight line,   I. 29 and 14.
    and *LE* and *EM* are in a straight line.

Find *GH*, a mean proportional between *BC* and *CF*,   VI. 13.
and on *GH* describe the rectilinear figure *KGH*, similar and
similarly situated to *ABC*.   VI. 18.
Then ∵ *BC* is to *GH* as *GH* is to *CF*,
    ∴ as *BC* is to *CF* so is *ABC* to *KGH*.   VI. 20, Cor. 2.
    But as *BC* is to *CF* so is ☐ *BE* to ☐ *EF*,   VI. 1,
and ∴ as *ABC* is to *KGH* so is ☐ *BE* to ☐ *EF*.   V. 5.
    Now *ABC* is equal to ☐ *BE*,   Constr.
and ∴ *KGH* = ☐ *EF*.   V. 14.
    But ☐ *EF* = the figure *D*.
    ∴ *KGH* = *D*; and *KGH* is similar to *ABC*.

Hence a figure *KGH* has been described as was required.

Q. E. F.

Def. V. A straight line is said to be cut in extreme and mean ratio, when the whole is to the greater segment as the greater segment is to the less.

PROPOSITION XXX. PROBLEM. (Eucl. VI. 30.)

To cut a straight line in extreme and mean ratio.

A————————————C————————B

Let AB be the given st. line.

It is required to cut AB in extreme and mean ratio.

Divide AB in the pt. C, so that rect. AB, BC = sq. on AC.

II. 11.

Then ∵ rect. AB, BC = sq. on AC.

∴ AB is to AC as AC is to BC, VI. 17.

and ∴ AB is cut in extreme and mean ratio in C. Def. 5.

Q. E. F.

Ex. 1. If two diagonals of a regular pentagon be drawn to cut one another, they cut one another in extreme and mean ratio.

Ex. 2. If the radius of a circle be cut in extreme and mean ratio, the greater segment will be equal to the side of a regular decagon described in the circle.

PROPOSITION XXXI. THEOREM.   (Eucl. VI. 32.)

If two triangles, SIMILARLY SITUATED, *which have two sides of the one proportional to two sides of the other, be joined at one angle, so as to have their homologous sides parallel, each to each, the remaining sides must be in a straight line.*

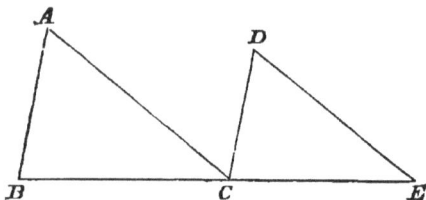

Let the △s *ABC*, *DCE* be similarly situated, having the sides *BA*, *AC* proportional to *CD*, *DE*, and let *BA* be ∥ to *CD*, and *AC* ∥ to *DE* ;

Then must *BC* and *CE* be in one st. line.

For ∵ *AC* meets the ∥s *BA*, *CD*,

∴ ∠ *BAC* = alternate ∠ *ACD*.   I. 29.

And ∵ *CD* meets the ∥s *AC*, *DE*,

∴ ∠ *ACD* = alternate ∠ *CDE*.   I. 29.

Hence ∠ *BAC* = ∠ *CDE*.   Ax. 1.

Then ∵ *BA* is to *AC* as *CD* is to *DE*, and ∠ *BAC* = ∠ *CDE*,

∴ △ *ABC* is equiangular to △ *DCE*.   VI. 6.

∴ ∠ *ACB* = ∠ *DEC* ;   VI. Def. 1.

and ∴ ∠ s *ACB*, *ACE* together = ∠ s *ACE*, *DEC* together,

= two right angles.   I. 29.

∴ *BC* and *CE* are in the same st. line.   I. 14.

Q. E. D.

1. Two common tangents to two circles meet at A. If the diameter of the smaller circle, the distance between the centres, and the diameter of the larger circle, be in the ratio of 1, 2, 3, prove that the distance from A to the centre of each circle is equal to the diameter of that circle.

2. Straight lines are drawn through the angular points of a triangle, parallel to the opposite sides, and through the angular points of the triangle thus formed straight lines are drawn, parallel to its opposite sides, and so on ; show that all these triangles are similar to the original triangle, and that any one of them has its sides bisected by the angular points of the preceding triangle.

3. If a point be taken within an equilateral triangle, the perpendiculars drawn from it to the three sides are together equal to the perpendicular drawn from one of the angles to the opposite side.

4. Upon AB as base two triangles ABC, ABD are described, and a line cutting CA is drawn parallel to CD. From the points where this line meets AC, AD, lines are drawn to meet CB, DB, and parallel to the base. Shew that these lines are equal.

5. If O be the centre, and AB the diameter of a circle, and if on AO as a diameter a circle be described, then the circumference of this circle will bisect any chord, drawn through it from A to meet the exterior circle.

6. On a given base describe a triangle, having a given vertical angle, and one of its sides double of the other.

7. From a point E in the common base of two triangles ACB, ADB, straight lines are drawn parallel to AC, AD, meeting BC, BD in F and G. Shew that the lines joining F, G and C, D will be parallel.

8. From the angular points, of a triangle ABC, straight lines AD, BE, CF, are drawn perpendicular to the opposite sides

and terminated by the circumscribing circle ; if L be the point of their intersection, shew that LD, LE, LF are bisected by the sides of the triangle.

9. If D and E be points in the sides of a triangle ABC, such that AD and AE are respectively the third parts of AB and AC, shew that BE and CD cut one another in a point of quadrisection.

10. In AB, AC, two sides of a triangle, are taken points D, E ; AB, AC are produced to F, G, such that $BF = AD$, and $CG = AE$: and BG, CF, FG are joined, the two former meeting in H. Show that the triangle FHG is equal to the triangles BHC, ADE together.

11. If the angle, between the internal bisector of the angle of a triangle and the base, be equal to the angle between the external bisector and the greater side produced, a perpendicular on this side through the vertex will bisect the segment of the base between the internal and external bisectors.

12. Triangles on equal bases and between the same parallels will have equal areas cut off by a line parallel to their bases.

13. From A, B, the extremities of the diameter of a circle, lines ACE, BCD, are drawn through a point C, on the circumference, to points E and D, such that EB and DA touch the circle. Shew that ED meets the tangent at C in AB produced.

14. Draw a straight line cutting two concentric circles, so that the part of it which is intercepted by the circumference of the greater may be four times as great as the part intercepted by the circumference of the less.

15. Shew how to inscribe a rectangle $DEFG$ in a triangle ABC, so that the angles D, E may be in AB, AC respectively, the side FG coincident with the base, and the area of the rectangle be equal to half that of the triangle.

16. If the bisectors of the opposite angles A, C, of a quadrilateral figure $ABCD$, intersect on the diagonal BD, then will the bisectors of the angles B, D meet on AC.

17. Two sides of a quadrilateral described about a circle are

parallel ; if the points of contact divide the other two sides proportionally, they are equally inclined to the first two.

18. If two triangles, on the same base, have their vertices joined by a straight line, which meets the base, or the base produced, shew that the parts of this line, between the vertices of the triangles and the base, are in the same ratio to each other as the areas of the triangles.

19. If perpendiculars be drawn from any point on the circumference of a circle to two tangents and the chord joining the points of contact, shew that the square on the perpendicular to the chord is equal to the rectangle contained by the other perpendiculars.

20. If the angles B, C, of the triangle ABC, be respectively equal to the angles D, E, of the triangle ADE, and the angles B, E, of the triangle ABE, to the angles D, C, of the triangle ADC, then these pairs of triangles shall be respectively equal to each other ; and if BE, CD, intersect in F, the triangles BFD, CFE, shall also be similar.

21. If, from the extremities of the diameter of a semicircle, perpendiculars be let fall on any line cutting the semicircle, the parts intercepted between those perpendiculars and the circumference are equal.

22. In a given circle place a chord, parallel to a given chord, and having a given ratio to it.

23. ABC is an equilateral triangle. Through C a line is drawn at right angles to AC, meeting AB produced in D, and a line through A parallel to BC in E. Through K, the middle point of AB, lines are drawn respectively parallel to AE, AC, and meeting DE in F and G. Prove that the sum of the squares on KG and FG is equal to three times the square on FE.

24. Find a point in the base of a right-angled triangle produced such that the line drawn from it to the angular point opposite to the base, shall be to the base produced as the perpendicular to the base itself.

25. AB is a given straight line, and D a given point in it; it is required to find a point P, in AB produced, such that AP is to PB as AD is to DB.

26. If two circles touch each other externally, and parallel diameters be drawn, the straight line, joining the extremities of those diameters, will pass through the point of contact.

27. If two circles touch each other, and also touch a straight line; the part of the line, between the points of contact, is a mean proportional between the diameters of the circles.

28. Two circles touch each other internally, the radius of one being treble that of the other. Shew that a point of tri-section of any chord of the larger circle, drawn from the point of contact, is its intersection with the circumference of the smaller circle.

29.. If ABC be a right-angled triangle, and D any point in its hypotenuse AB, determine by a geometrical construction the point P, to which AB must be produced, so that PA is to PB as AD is to DB.

30. If a line touching two circles cut another line joining their centres, the segments of the latter will be to each other as the diameters of the circles.

31. If through the vertex of an equilateral triangle a perpendicular be drawn to the side, meeting a perpendicular to the base, drawn from its extremity, the line, intercepted between the vertex and the latter perpendicular, is equal to the radius of the circumscribing circle.

32. If on the diagonals of a quadrilateral as bases, parallelo-grams be described, equal to the quadrilateral, find the ratio of their altitudes.

33. The opposite sides AB, DC of a quadrilateral $ABCD$, which can be inscribed in a circle, meet, when produced, at E; F is the point of intersection of the diagonals, and EF meets AD in G; prove that the rectangle EA, AB is to the rectangle ED, DC as AG is to GD.

34. If from the extremities of the diameter of a circle tangents be drawn, any other tangent of the circle, terminated by them, is so divided at its point of contact, that the radius of the circle is a mean proportional between the segments of the tangent.

35. If the sides of a triangle, inscribed in the segment of a circle, be produced to meet lines drawn from the extremities of the base, forming with it angles equal to the angle in the segment, the rectangle contained by these lines will be equal to the square on the base.

36. Describe a parallelogram, which shall be of a given altitude, and equal and equiangular to a given parallelogram.

37. Two circles touch each other internally at the point A, and from two points in the line joining their centres perpendiculars are drawn, intersecting the outer circle in the points B, C, and the inner circle in the points D, E. Shew that AB is to AC as AD is to AE.

38. Given of any triangle the base, and the point, where the line, bisecting the exterior vertical angle, cuts the base produced, find the locus of the vertex of the triangle.

39. Draw a line from one of the angles at the base of a triangle, so that the part of it cut off by a line drawn from the vertex parallel to the base, may have a given ratio to the part cut off by the opposite side.

40. If AC be drawn from A to a point C in the base of the triangle ABD, so that ABD, ACD are similar triangles, shew that DA touches the circle described about ABC.

41. If the centres A, B, of two circles be joined, and P be the point in the line AB, from which equal tangents can be drawn to the circles ; the tangents drawn from any point in a line, which passes through P at right angles to AB are all equal.

42. Construct a triangle, similar to a given triangle, and having its angular points upon three given straight lines, which meet in a point.

43. Let $ABCD$ be any parallelogram, BD its diagonal. Then the perpendiculars, from A on BD, and from B and D upon AD and AB, shall all pass through a point.

44. If a quadrilateral be inscribed in a circle, its diagonals shall be to one another as the sums of the rectangles contained by the sides adjacent to their extremities.

45. A square is described on the base of an isosceles triangle, remote from the vertex. Prove that, if the vertex be joined to the corners of the square, the middle segment of the base will be to the outer one in twice the ratio of the perpendicular on the base to the base.

46. The base AB of an isosceles triangle ABC is produced both ways to D and E, so that the rectangle AD, BE is equal to the square on AC. Shew that the triangles DAC, EBC, are similar.

47. If each of the angles at the base of an isosceles triangle be double of the angle at the vertex, shew that either side is a mean proportional between the perimeter of the triangle, and the distance of the centre of the inscribed circle from either end of the base.

48. ABC is a triangle, and O is the centre of the circle inscribed in the triangle. Shew that AO passes through the centre of the circle described about the triangle BOC.

49. Draw a line parallel to one of the sides of a triangle, so that it may be a mean proportional between the segments into which it divides one of the other sides.

50. If an equilateral triangle be inscribed in a circle, and the adjacent arcs cut off by two of its sides be bisected, shew that the line joining the points of bisection will be trisected by the sides.

51. ABC is an equilateral triangle, BC is produced to D, and CD is made equal to BC: CE is drawn at right angles to DCB, and at A the angle CAE is made equal to the angle DCA; DE, DA are drawn. Shew that the rectangle DA,

CE is equal to the rectangle *DE, AC* together with the square on *CB*.

52. Two straight lines *AB, CD*, intersect in *E*. If when *AC, BD* are joined, the sides of the triangle *ACE*, taken in order, are proportional to those of the triangle *DBE*, taken in order, shew that *A, C, B, D*, lie on the circumference of the same circle.

53. If any triangle be inscribed in a circle, and from the vertex a line be drawn parallel to a tangent at either extremity of the base, this line will be a fourth proportional to the base and two sides.

54. If a triangle be inscribed in a semicircle, and a perpendicular be drawn from any point in the diameter, meeting one side, the circumference, and the other side produced : the segments cut off will be in continued proportion.

55. If *ABCD* be any quadrilateral figure inscribed in a circle, and *BK, DL* be perpendiculars on the diagonal *AC*, shew that *BK* is to *DL* as the rectangle *AB, BC* is to the rectangle *AD, DC*.

56. If a rectangular parallelogram be inscribed in a right-angled triangle, and they have the right-angle common, the rectangle, contained by the segments of the hypotenuse, is equal to the sum of the rectangles, contained by the segments of the sides about the right angle.

57. If from the vertex of an isosceles triangle a circle be described, with a radius less than one of the equal sides, but greater than the perpendicular from the vertex to the base, the parts of the base cut off by it will be equal.

58. Through a fixed point *A* on a circle, a chord *AB* is drawn, and produced to a point *M*, so that the rectangle contained by *AB* and *AM* is constant. Find the locus of *M*.

59. If two sides of a triangle be unequal, the sum of the greater side and the perpendicular upon it from the opposite angle is greater than the sum of the less side and the perpendicular upon it from the opposite angle.

60. From one angle of a triangle, perpendiculars are dropped on the external bisectors of the other two angles ; prove that the distance between the feet of these perpendiculars is equal to half the sum of the sides of the triangle.

61. A, B, P, Q, R, are five points in the circumference of a circle ; p, q, r, are the intersections of perpendiculars of the triangles ABP, ABQ, ABR respectively ; prove that the triangles PQR, pqr are similar, equal, and similarly placed.

62. AD, BE, CF are perpendiculars from the angular points of a triangle on the opposite sides, intersecting in P. Prove that the rectangle AP, BC is equal to the sum of the rectangles PE, AC and PF, AB.

63. ABC is a triangle, and AD, AE, are drawn to points D, E, in the base, so as to make equal angles with AB, AC, respectively. Shew that the square on AB is to the square on AC as the rectangle BD, BE is to the rectangle CD, CE.

64. Find a straight line, such that the perpendiculars, let fall upon it from three given points, shall be in a given ratio to each other.

65. Find a fourth proportional to three given similar triangles.

66. If the sides of a triangle be bisected, and the points joined with the opposite angles, the joining lines shall divide each other proportionally, and the triangle, formed by the joining lines, and the remaining side, shall be equal to a third of the original triangle.

67. Find the locus of a point, such that the distance between the feet of the perpendiculars from it upon two straight lines, given in position, may be constant.

68. $ABCD$ is a parallelogram, AC, BD diagonals. If parallel lines be drawn through A, C, and also through B, D, the diagonals of all parallelograms so formed will pass through the same point.

69. OPQ is any triangle. OR bisects PQ in R ; PST bisects OR in S, and cuts OQ in T. Shew that $OQ=3OT$.

70. If the side BC, of a triangle ABC, be bisected by a line, which meets AD and AC, produced if necessary, in D and E respectively, shew that AE is to EC as AD is to DB.

71. Two circles are drawn in the same plane, having a common centre C. If the tangent, at any point P of the inner circle, meet the outer in Q, and be produced both ways to points A, B, such that QA, QB, are each of them equal to QC, the area of the triangle CAB will be constant.

72. From P, a point without a circle, whose centre is C, two tangents PA, PB, are drawn, and also a line, meeting the circle in D, and AB in E. If CF be perpendicular to PD, then FD is a mean proportional between FP and FE.

73. Three circles touch the sides of a triangle ABC in the points where the inscribed circle touches them, and touch each other, in the points G, H, K. Prove that AG, BH and CK meet in a point.

74. If ABC be a right-angled triangle, and EF, parallel to BC, the hypotenuse, meet AB, AC in E, F, then EH, FL, AK being drawn perpendicular to BC, shew that the difference of the rectangles CK, CH and BL, BK is equal to the difference of the squares on AB, AC.

75. From a point A in the circumference of a circle two chords AB, AC are drawn, cutting off arcs greater than a quadrant and less than a semicircle ; and from the extremity B of the greater chord, a line BD is drawn in a direction perpendicular to that of the diameter through A, and meets AC produced in D. Shew that AD is to AB as AB is to AC.

76. Two circles intersect, and through a point of intersection two lines are drawn, terminated by the circumferences of both circles ; one of these lines remains fixed, while the other may have any position. Shew that the locus of the intersection of the lines joining their extremities is a circle.

77. If the side BC of an equilateral triangle ABC be produced to any point D, and AD be joined, and if a straight line CE be drawn parallel to AB, cutting AD in E, prove that the square on AE is to the rect. DA, DE as the rect. CE, CB is to the square on DC.

78. In a triangle, right-angled at A, if the side AC be double of AB, the angle B is more than double the angle C.

79. From the obtuse angle of a triangle draw a line to the base, which shall be a mean proportional between the segments, into which it divides the base.

80. AB, AC are two straight lines, B and C given points in the same; BD is drawn perpendicular to AC, and DE perpendicular to AB; in like manner CF is drawn perpendicular to AB, and FG to AC. Shew that EG is parallel to BC.

81. AB is the diameter of a circle, and CD a chord at right angles to it, E any point in CD. If AE and BE be drawn and produced to cut the circle in F and G, the quadrilateral $FCGD$ has any two of its adjacent sides in the same ratio as the remaining two.

82. $ADEB$ is a semicircle; AB the diameter; DF, EG perpendiculars on the diameter; C the centre of a circle, which touches the semicircle and these perpendiculars; and CH is drawn perpendicular to the diameter. Shew that CH is a mean proportional between AF and BG.

83. Divide a straight line in a given ratio, and produce it so that the whole line thus produced shall be to the part produced in the same ratio; shew that the circle described on the line between the two points of section, as diameter, is such, that if any point of its circumference be joined with the extremities of the given line, the straight lines so drawn shall also be in the given ratio.

84. If any secant be drawn through the intersection of two tangents to a circle, and if the points of intersection be joined to the points of contact of the tangents, shew that the rectangles under the pairs of opposite sides of the quadrilateral formed by the joining lines are equal.

85. Triangles on the same base, and with equal vertical angles, are to one another as the products of their sides.

86. A line $ACBD$ is divided, so that AC is to CB as AD is to DB. Shew that a semicircle, described on CD, is the locus of P, such that AP is to PB as AC is to CB.

87. If the two diagonals of a quadrilateral, inscribed in a circle, be given, shew that the quadrilateral is greatest, when they are at right angles.

88. ABC is a triangle, D, E, the middle points of AB, AC, and BE, CD, meet in F : a triangle is drawn, having its sides parallel to AF, BF, CF. Shew that the lines, joining its angular points to the middle points of its opposite sides, will be parallel to the sides of the triangle ABC.

89. A circle rolls within another, of twice its radius : if P be the point of contact, and A a given point of the rolling circle, PA will be constant in direction.

90. Two circles intersect ; the line $AHKB$ joining their centres A, B, meets them in H, K. On AB is described an equilateral triangle ABC, whose sides BC, AC intersect the circles in F, E. FE produced meets BA produced in G. Shew that as GA is to GK, so is CF to CE, and so also is GH to GB.

91. ABC is a triangle inscribed in a circle, and perpendiculars are drawn from any point in the circumference to the sides of the triangle. Prove that the points in which they meet the sides are in one straight line.

92. An isosceles triangle has one of its equal sides a mean proportional between two sides of another triangle. If these two sides include the same angle as the vertical angle of the isosceles triangle, shew that the triangles are equal.

93. Two triangles ABC, BCD, have the side BC common, the angles at B equal, and the angles ACB, BDC right angles. Shew that the triangle ABC is to the triangle BCD as AB is to BD.

94. Given the straight line which is drawn from the vertex of an equilateral triangle to a point of trisection of the base, find the side of the triangle.

95. Straight lines being drawn from the angular points A, B, C, of a triangle through any the same point, so as to cut the opposite sides respectively in a, b, c, shew that the rectangle Ab, Bc is to the rectangle Ac, Ba as Cb is to Ca.

96. *ABCD* is a quadrilateral inscribed in a circle, and its diagonals intersect in *F*. Shew that the rectangle *AF*, *FD* is to the rectangle *BF*, *FC* as the square on *AD* is to the square on *BC*.

97. *ABCD* is a quadrilateral figure whose opposite angles are not supplemental; the circle described about *ABD* cuts *DC* in *E*, and the circle described about *BCE* cuts *AE* in *F*. Shew that the triangle *ABF* is equiangular to the triangle *BCD*, and the triangle *BCF* to the triangle *ABD*.

98. *ACB* is a triangle whereof the side *AC* is produced to *D* until *CD* is equal to *AC*; and *BD* is joined: shew that if any line drawn parallel to *AB* cuts the sides *AC* and *CB*, and from the points of section lines be drawn parallel to *DB*, these will meet *AB* in points equidistant from its extremities.

99. *A* and *B* are fixed points, and *AC*, *BD* are perpendiculars on *CD*, a given straight line : the straight lines *AD*, *BC*, intersect in *E*, and *EF* is drawn perpendicular to *CD*. Shew that *EF* bisects the angle *AFB*.

100. If *O* be the centre of a circle circumscribed about the triangle *ABC*, obtuse-angled at *C*, and if on *OC* as diameter a circle be described meeting *AB* in *D* and *E*, then either *CD* or *CE* shall be a mean proportional between the segments into which they respectively divide *AB*.

101. The exterior angle *CBD* of the triangle *ABC* is bisected by the line *BE*, which cuts the base produced in *E*. Shew that the square on *BE*, together with the rectangle *AB*, *BC*, is equal to the rectangle *AE*, *EC*.

102. *ABCD* is a quadrilateral figure inscribed in a circle; *BA*, *CD*, are produced to meet in *P*, and *AD*, *BC*, are produced to meet in *Q*. Prove that *PC* is to *PB* as *QA* is to *QB*.

Also, shew that half the sum of the angles at *P* and *Q* is equal to the complement of the opposite angle *ABC* of the quadrilateral figure.

103. Having given the vertical angle, and the ratio of the sides containing it, and also the diameter of the circumscribing circle, construct the triangle.

104. From the centre of a given circle draw a straight line to meet a given tangent to the circle, so that the segment of the line between the circle and the tangent shall be any required part of the tangent.

105. Find a point the distances of which from three given points not in the same straight line are proportional to p, q, r respectively, the four points being in the same plane.

106. AB is the diameter of a circle, D any point in the circumference, and C the middle point of the arc AD. If AC, AD, BC, be joined, and AD cut BC in E, the circle described about the triangle AEB will touch AC, and its diameter will be a third proportional to BC and AB.

107. From a given point A a variable straight line is drawn, meeting a fixed straight line in P, and a point Q is taken on it so that the rectangle AP, AQ is constant. Find the locus of Q.

108. On a given base describe a rectangle, which shall be equal to the difference of the squares on two given straight lines, any two of the three given lines being together greater than the third.

109. If the exterior angles of a triangle be bisected by straight lines, forming another triangle, shew that the two triangles cannot be similar, unless they be each equilateral.

110. If ABC, $A'B'C'$ be similar triangles, and $AB = A'C'$, shew the areas of the triangles are as AC to $A'B'$.

111. The alternate angles of a regular hexagon are joined: shew that the area of the hexagon formed by the intersections of the joining lines is one-third of the original hexagon.

112. A triangle is divided by a straight line parallel to the base into two parts, the areas of which are as 1 to 8 : how does the straight line divide the sides ?

113. The line AD is divided into three equal parts in the points B and C; a circle is described with B as centre and BA as radius, and any circle cutting this is described with D as centre. Shew that if a chord to both the circles be drawn

21

from A, through one of the points of intersection, it will be bisected by this point.

114. ABC is an acute-angled triangle, E and F are the middle points of the sides AB and AC. Shew that a line drawn from E, equal to EA, to meet the base, and another from F, equal to FA, also to meet it, will intersect the base at the same point.

Hence explain how, by folding a piece of paper such as the triangle ABC, it may be shewn that the three angles of a triangle are equal to two right angles.

115. If ABC, ADE be two equal triangles having the angles BAC, DAE equal, and if they be placed so that BA, AE are in a straight line, as also CA and AD; and if BC, DE be produced to meet in F, prove that FA will bisect CE and also BD.

116. Within a circle, whose diameter is AB, another circle is inscribed, touching the outer circle in A, and passing through its centre O. From a point N, in AB, a line NQP is drawn perpendicular to AB, meeting the inner circle in Q, and the outer circle in P, AN being equal to one-sixth of AB. Prove that the duplicate ratio of NQ to NP is equal to the ratio of 2 to 5.

117. Describe a square, which shall be equal to the sum of a given square and a given rectangle, a side of the given square being greater than half the difference of the two sides containing the rectangle.

BOOK XI.

INTRODUCTORY REMARKS.

In Book I. Def. 7., it is laid down that a Plane Surface is one in which, if any two points be taken, the straight line between them lies wholly in that surface.

This definition should be extended by the addition of the following words, *and if the straight line be produced, every point in the part produced will lie in the plane.*

Euclid professes to prove this in the first Proposition of Book XI., which is thus enunciated : "one part of a straight line cannot be in a plane, and another part out of the plane."

But this has been assumed again and again in the proofs of earlier propositions ; thus, for example, we have called a circle a *plane figure*, and having drawn any radius to a circle we have assumed that the radius, produced within the circumference, will meet the circumference.

From the extended definition of a Plane Surface it follows that a straight line, which meets a plane, must either lie entirely in that plane, or meet it in *one* point only ; for if it met the plane in *two* points, it would lie entirely in the plane.

The Definitions given at the commencement of Book XI. relate partly to Plane Surfaces and partly to Solid Figures. By a slight change in the order in which they stand in the Greek text, we obtain the advantage of arranging them in accordance with this twofold division.

DEFINITIONS.

Relating to Plane Surfaces.

I. A Plane Surface is one in which, if any two points be taken, the straight line between them lies wholly in that surface ; and if the straight line be produced, every point in the part produced will lie in the plane.

II. When a straight line is at right angles to *every* straight line in a plane which meets it, it is said to be perpendicular to the plane.

Note.— It will be shown in Prop. IV. that when a straight line is at right angles to each of two other straight lines in a plane, which meet it, it is at right angles to every other straight line in the plane which meets it.

III. A plane is perpendicular to a plane, when the straight lines, drawn in one of the planes perpendicular to the common section of the two planes, are perpendicular to the other plane.

IV. The inclination of a straight line to a plane is the acute angle, contained by that straight line and another, drawn from the point at which the first line meets the plane, to the point at which a perpendicular to the plane, drawn from any point of the first line above the plane, meets the same plane.

V. The inclination of a plane to a plane is the acute angle, contained by two straight lines, drawn from any the same point of their common section, at right angles to it, one in one plane, and the other in the other plane.

VI. Two planes are said to have the same inclination to one another, which two other planes have, when the said angles of inclination are equal to one another.

VII. Parallel planes are such as do not meet one another though produced.

Relating to Solid Figures.

VIII. A Solid is that which has length, breadth, and thickness.

IX. That which bounds a solid is a superficies.

X. A Solid Angle is that, which is made by the meeting of more than two plane angles, which are not in the same plane, at one point.

Definitions I. to X. are all that are required in the part of Book xi. included in this work. Those which follow are necessary to the explanation of some of the terms, which will be found in the Exercises and Examination Papers.

XI. Similar solid figures are such, as have all their solid angles equal, each to each, and are contained by the same number of similar planes.

XII. A Pyramid is a solid figure, contained by planes, which are constructed between one plane and one point above it, at which they meet.

XIII. A Prism is a solid figure, contained by plane figures, of which two that are opposite are equal, similar, and parallel to one another ; and the others are parallelograms.

XIV. A Sphere is a solid figure, described by the revolution of a semicircle about its diameter, which remains fixed.

XV. The Axis of a Sphere is the fixed straight line, about which the semicircle revolves.

XVI. The Centre of a Sphere is the same with that of the semicircle.

XVII. The Diameter of a Sphere is any straight line, which passes through the centre, and is terminated both ways by the superficies of the sphere.

XVIII. A Cone is a solid figure, described by the revolution of a right-angled triangle about one of the sides containing the right angle, which side remains fixed. If the fixed side be equal to the other side containing the right angle, the cone is called a right-angled cone ; if it be less than the other side, an obtuse-angled cone ; and if greater, an acute-angled cone.

XIX. The Axis of a Cone is the fixed straight line, about which the triangle revolves.

XX. The Base of a Cone is the circle, described by that side, containing the right angle, which revolves.

XXI. A Cylinder is a solid figure, described by the revolution of a rectangle about one of its sides, which remains fixed.

XXII. The Axis of a Cylinder is the fixed straight line about which the rectangle revolves.

XXIII. The Bases of a Cylinder are the circles, described by the two revolving opposite sides of the rectangle.

XXIV. Similar cones and cylinders are those which have their axes and the diameters of their bases proportionals.

XXV. A Cube is a solid figure, contained by six equal squares.

XXVI. A Tetrahedron is a solid figure, contained by four equal and equilateral triangles.

XXVII. An Octahedron is a solid figure, contained by eight equal and equilateral triangles.

XXVIII. A Dodecahedron is a solid figure, contained by twelve equal pentagons, which are equilateral and equiangular.

XXIX. An Icosahedron is a solid figure, contained by twenty equal and equilateral triangles.

XXX. A Parallelepiped is a solid figure, contained by six quadrilateral figures, of which every opposite two are parallel.

POSTULATE.

Let it be granted that a plane may be made to pass through any given straight line.

PROPOSITION I.   THEOREM.   (Eucl. XI. 2.)

If two straight lines meet one another, a plane can be drawn to contain both ; and every plane containing both must coincide with the aforesaid plane.

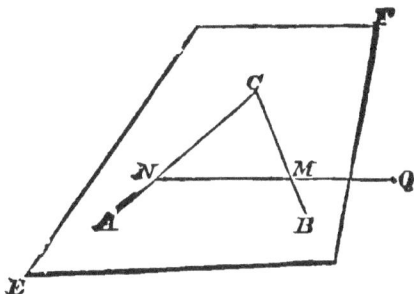

Let the two st. lines AC, BC meet in C.

Then a plane can be drawn to contain AC and BC.

Let any plane EF be drawn to contain AC,   Post.

and let EF be turned about AC till it pass through B.

Then ∵ B and C are points in the plane EF,

∴ BC lies in the plane EF.   XI. Def. 1.

Also, any plane containing AC and BC must coincide with EF.

For let Q be any point in a plane containing AC and BC.

Draw QMN in this plane to cut BC, AC in M and N.

Then ∵ M and N are points in the plane EF,

∴ Q is a point in the plane EF.   XI. Def. 1.

Similarly, any point in a plane containing AC, BC must lie in EF ;

and ∴ any plane containing AC, BC must coincide with EF.

Q. E. D.

COR. I. *Hence it follows that a plane is completely determined by the condition that it passes through two intersecting straight lines.*

Cor. II. *A straight line and a point without the line determine
a plane.*

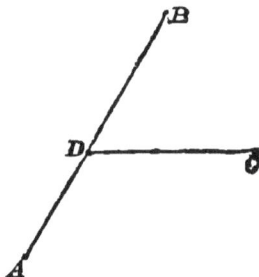

Let AB be a straight line, and C a point without AB.

Draw the st. line CD to any point D in AB.

Then one plane can be drawn to contain AB and CD.   XI. 1,

∴ one...AB and C.

Again, any plane containing AB must contain D,

∴ any plane containing AB and C must contain CD also.

But there is only one plane that can contain AB and CD,

∴ there is only one plane ....AB and C.

Hence the plane is completely determined.

Cor. III. *Three points, not in the same straight line, determine
a plane.*

For let A, B, C be three such points (fig. Cor. 2).

Draw the straight line AB.

Then a plane, which contains A, B and C, must contain AB
and C,

and a plane, which contains AB and C, must contain A, B, C.

Now AB and C are contained by one plane, and one only,

Cor. 2.

∴ A, B, C are contained by one plane, and one only.

Hence the plane is completely determined.

Cor. IV. *Two parallel lines determine a plane.*

For, by the definition of parallel lines, the two lines are in
the same plane, and as only one plane can be drawn to contain
one of the lines and any point in the other line, it follows that
only one plane can be drawn to contain both lines.

If two planes cut one another, their common section must be a straight line.

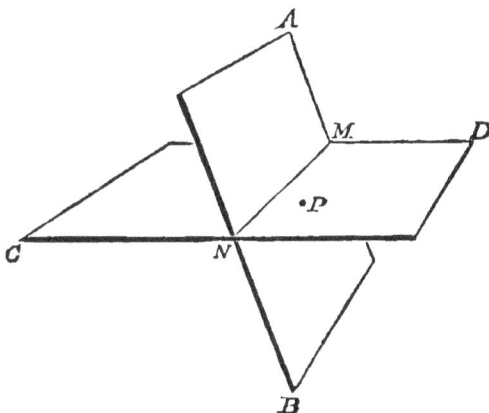

Let *AB* and *CD* be two planes that cut one another.

Then must their common section be a straight line.

Let *M* and *N* be two points common to both planes.
Draw the straight line *MN*.
Then ∵ *M* and *N* are common to both planes,
∴ the st. line *MN* lies in both planes.   XI. Def. 1.
And no point, out of this line, can be common to both planes.
For, if it be possible, let *P* be such a point.
But there can be but *one* plane common to the point *P* and
the st. line *MN*.   XI. 1, Cor. 2.
∴ *P* is not common to *both* planes.

Hence every point in the common section of the planes lies
in the straight line *MN*.

Q. E. D.

Note.—The Propositions which follow are numbered as in
Euclid.

PROPOSITION IV. THEOREM.

If a straight line stand at right angles to each of two straight lines, at the point of their intersection, it must also be at right angles to the plane that passes through them.

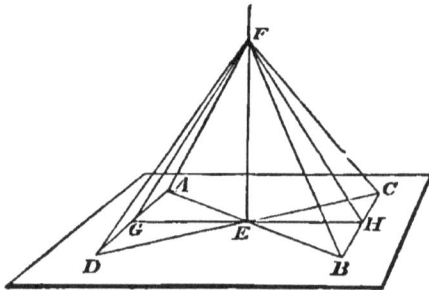

Let the st. line *EF* be ⊥ to each of the st. lines *AB, CD,* at *E,* the pt. of their intersection.

Then must EF be ⊥ to the plane passing through AB, CD.

Make *AE, EB, CE, ED,* all equal to one another, and through *E,* draw, in the plane in which *AB, CD* are, any st. line *GEH,* and join *AD, CB.*

Take any pt. *F,* in *EF,* and join *FA, FG, FD, FC, FH, FB.*

Then in △s *AED, BEC,*

∵ *AE=BE,* and *DE=CE,* and ∠ *AED* = ∠ *BEC,* I. 15.

∴ *AD=BC,* and ∠ *DAE* = ∠ *CBE,*   I. 4.

Then in △s *AEG, BEH,*

∵ ∠ *AEG* = ∠ *BEH,* and ∠ *GAE* = ∠ *HBE,* and *AE=BE,*

∴ *GE=HE,* and *AG=BH.*   I. B. p. 17.

Then in △s *AEF, BEF,*

∵ *AE=BE,* and *EF* is common, and rt. ∠ *AEF* = rt. ∠ *BEF,*

∴ *AF=BF.*   I. 4.

So also, $CF = DF$

Then in $\triangle$s ADF, BCF,

$\because AD = BC$, and $AF = BF$, and $DF = CF$

$\therefore \angle DAF = \angle CBF.$ I. c. p. 18.

Again, in $\triangle$s AFG, BFH,

$\because AF = BF$, and $AG = BH$, and $\angle FAG = \angle FBH$,

$\therefore FG = FH.$ I. 4.

Then in $\triangle$s FEG, FEH,

$\because GE = HE$, and EF is common, and $FG = FH$,

$\therefore \angle FEG = \angle FEH.$ I. c.

$\therefore EF$ is $\perp$ to GH.

In like manner it may be shown that EF is $\perp$ to every st. line which meets it in the plane passing through AB, CD.

$\therefore EF$ is $\perp$ to the plane, in which AB, CD are. XI. Def. 2.

Q. E. D

PROPOSITION V. THEOREM.

If three straight lines meet all at one point, and a straight line stand at right angles to each of them at that point, the c..ve straight lines must be in one and the same plane.

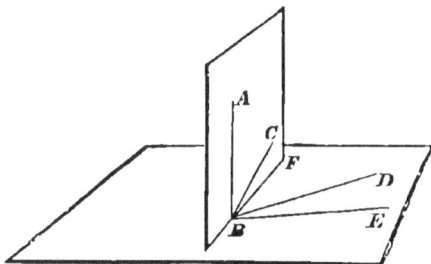

Let the st. line AB be $\perp$ to each of the st. lines BC, BD, BE, at B, the pt. where they meet.

Then must BC, BD, BE be in one ana the same plane.

If not, let BD, BE be in one plane, and BC without it, and let a plane, passing through AB, BC, cut the plane, in which BD and BE are, in the st. line BF.　　　　XI. 2.

Then AB, BC, BF are all in one plane.

And ∵ AB is $\perp$ to BD and BE,

　　∴ AB is $\perp$ to the plane in which BD and BE are, XI. 4.

and ∴ AB is $\perp$ to BF, a st. line in that plane.　　XI. Def. 2.

Thus ∠ ABF is a rt. ∠,

　　and ∠ ABC is a rt. ∠ ;　　　　　　　　　　　　　　Hyp.

　　　　　∴ ∠ $ABC = $ ∠ ABF,

the less = the greater, which is impossible.

　　∴ BC is not without the plane, in which BD, BE are,

and ∴ BC, BD, BE are in one and the same plane.

Q. E. D.

PROPOSITION VI.   THEOREM.

If two straight lines be at right angles to the same plane, they must be parallel to one another.

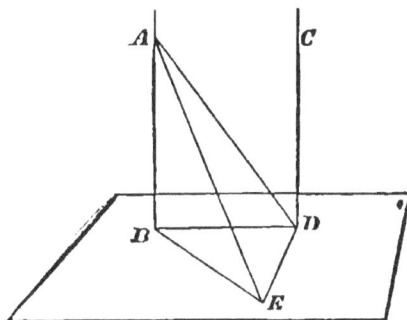

Let the st. lines AB, CD be $\perp$ to the same plane.

Then must AB be $\parallel$ to CD.

Let AB, CD meet the plane in the pts. B, D.

Join BD, and draw $DE \perp$ to BD, in the same plane.   I. 11

    Make $DE = AB$, and join BE, AE, AD.

Then $\because$ AB is $\perp$ to the plane,

    $\therefore$ AB is $\perp$ to BD and BE,   XI. Def. 2.

and $\therefore$ each of the $\angle$ s ABD, ABE is a rt. $\angle$.

So also, each of the $\angle$ s CDB, CDE is a rt. $\angle$.

Then, in $\triangle$ s ABD, EDB,

$\because$ $AB = ED$, and BD is common, and rt. $\angle$ $ABD =$ rt. $\angle$ EDB.

        $\therefore$ $DA = BE$.   I. 4.

Again, in $\triangle$ s ABE, EDA,

    $\because$ $AB = ED$, and $BE = DA$, and AE is common,

        $\therefore$ $\angle$ $ABE = \angle$ EDA.   I. c.

But $\angle$ ABE is a rt. $\angle$;

        $\therefore$ $\angle$ EDA is a rt. $\angle$.

and $\therefore$ ED is $\perp$ to AD.

Thus ED is $\perp$ to BD, AD, CD, at the pt. where they meet,

    and $\therefore$ BD, AD, CD are all in one plane.   XI. 5.

But AB is in the plane, in which BD and AD are ;   XI. 1.

    and $\therefore$ AB, BD, CD are all in one plane.

Then $\because$ each of the $\angle$ s ABD, CDB is a rt. $\angle$,

    $\therefore$ AB is $\parallel$ to CD.   I. 28.

Q. E. D.

Proposition VII. Theorem.

If two straight lines be parallel, the straight line drawn from any point in the one to any point in the other, is in the same plane with the parallels.

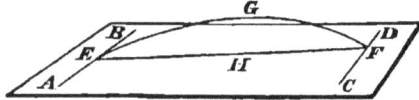

Let *AB* and *CD* be parallel straight lines.
Take any pts. *E, F* in *AB* and *CD*.

Then must the st. line joining E and F be in the same plane as
AB, CD.

If not, let it be without the plane, as *EGF*.

In the plane *ABCD*, in which the parallels are,
    draw the st. line *EHF* from *E* to *F*.

Then the two st. lines *EGF*, *EHF* enclose a space,
which is impossible.   I. Post. 5.

∴ the st. line joining *E* and *F* cannot be out of the plane,
in which the parallels *AB, CD* are.

∴ it is in that plane.

Q. E. D.

Note.—We have proved this Proposition as Cor. 17. to
Prop. 1.

If two straight lines be parallel, and one of them be at right angles to a plane, the other must be at right angles to the same plane.

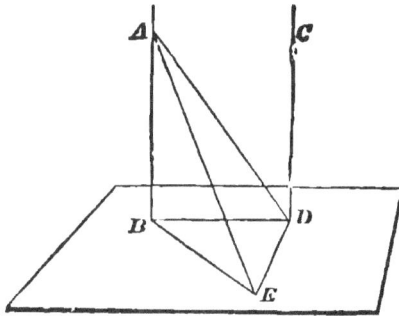

Let AB, CD be two ‖ st. lines,
and let one of them, AB, be ⊥ to a plane.

Then must CD be ⊥ to the same plane.

Let AB, CD meet the plane in the pts. B, D ; and join **BD**;
then AB, BD, CD are all in one plane.   XI. 7.
In the plane, to which AB is ⊥, draw DE ⊥ to BD,
make $DE = AB$, and join BE, AE, AD.
Then ∵ AB is ⊥ to the plane,
     ∴ each of the ∠ s ABD, ABE is a rt. ∠ ;   XI. Def. 2.
and ∵ BD meets the ‖ st. lines AB, CD,
     ∴ ∠ s ABD, CDB together = two rt. ∠ s,   I. 29.
and ∴ ∠ CDB is a rt. ∠ , and CD is ⊥ to BD.
Then in the △ s ABD, EDB,
∵ $AB = ED$, and BD is common, and rt. ∠ ABD = rt. ∠ EDB.
          ∴ $AD = EB$.   I. 4.
Then in △ s ABE, EDA,
     ∵ $AB = ED$, and AE is common, and $EB = AD$.
         ∴ ∠ ABE = ∠ EDA ;   I. c.
and ∴ ∠ EDA is a rt. ∠ .
Hence ED is ⊥ to DA, and it is also ⊥ to BD, by constr.,
     ∴ ED is ⊥ to the plane in which DA, BD are,   XI. 4.
and ∴ ED is ⊥ to DC, which is in that plane.   XI. Def. 2.

Hence CD is $\perp$ to DE.

Now CD is $\perp$ to DB.

∴ CD is $\perp$ to the plane passing through DE, DB. XI.4.

∴ CD is $\perp$ to the plane to which AB is $\perp$.

Q. E. D.

PROPOSITION IX. THEOREM.

Two straight lines, which are each of them parallel to the same straight line, and not in the same plane with it, are parallel to one another.

Let AB, CD be each of them $\parallel$ to EF,

and not in the same plane with it.

Then must AB be $\parallel$ to CD.

In EF take any pt. G.

From G draw, in the plane $ABEF$, $GH \perp$ to EF,

and, in the plane $CDEF$, $GK \perp$ to EF.   I. 11

Then ∵ EF is $\perp$ to GH and GK,

∴ EF is $\perp$ to the plane HGK;   XI. 4,

and ∵ EF is $\parallel$ to AB,

∴ AB is $\perp$ to the plane HGK.   XI. 8.

So also CD is $\perp$ to the plane HGK.   XI. 8.

∴ AB is $\parallel$ to CD.   XI. 6.

Q. E. D.

PROPOSITION X. THEOREM.

If two straight lines meeting one another be parallel to two others, that meet one another, and are not in the same plane with the first two, the first two and the other two must contain equal angles.

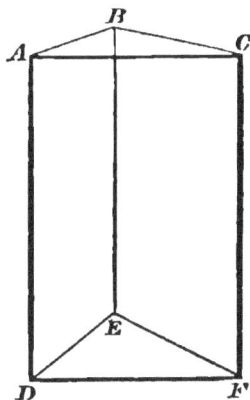

Let the two st. lines AB, BC, meeting at B in the plane ABC, be ∥ to the st. lines DE, EF, meeting at E in the plane DEF.

Then must ∠ ABC = ∠ DEF.

Make $BA = ED$, and $BC = EF$,   I. 3.

   and join AD, BE, CF, AC, DF.

Then ∵ AB is = and ∥ to DE,

   ∴ AD is = and ∥ to BE.   I. 33.

So also, CF is = and ∥ to BE.

   ∴ AD is = and ∥ to CF,   **Ax.** 1 and XI. 9.

and ∴ AC is = and ∥ to DF.   I. 33.

Then in △s ABC, DEF

   ∵ $AB = DE$, and $BC = EF$, and $AC = DF$,

   ∴ ∠ ABC = ∠ DEF.   I. c.

Q. E. D.

PROPOSITION XI. PROBLEM.

To draw a straight line perpendicular to a given plane, from a given point without it.

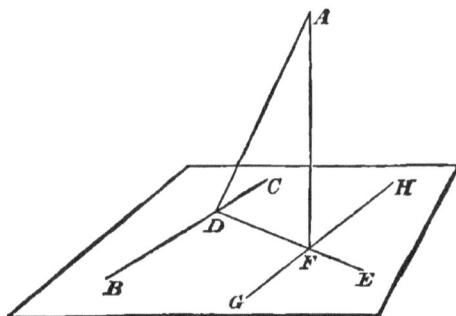

Let A be the given pt. without the plane BH.

It is required to draw from A a st. line $\perp$ to the plane BH.

In the plane, draw any st. line BC,
    and from A draw $AD \perp$ to BC.   I. 12.
Then if AD be $\perp$ to the plane, what was required is done.
If not, from D draw, in the plane BH, $DF \perp$ to BC.   I. 11.
    and from A draw $AF \perp$ to DE :   I. 12.
    AF will be $\perp$ to the plane BH.
Through F, draw $GH \parallel$ to BC.   I. 31.
Then $\because$ BC is $\perp$ to both AD and DE,
    $\therefore$ BC is $\perp$ to the plane AFD ;   XI. 4.
    and GH is $\parallel$ to BC,
    $\therefore$ GH is $\perp$ to the plane AFD.   XI. 8.
Hence GH is $\perp$ to the line AF in that plane ;   XI. Def. 2.
           and $\therefore$ AF is $\perp$ to GH.
Also, AF is $\perp$ to DE, by construction ;
    $\therefore$ AF is $\perp$ to the plane passing through GH, DE,   XI.4.
that is, AF is $\perp$ to the plane BH.
    Thus from A a line AF is drawn $\perp$ to the plane BH.

Q. E. F.

To erect a straight line at right angles to a given plane, from a given point in the plane.

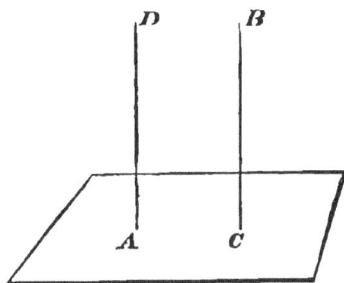

Let A be the given pt. in the given plane.

It is required to erect a st. line from $A \perp$ to the plane.

From any pt. B, without the plane, draw $BC \perp$ to it,   XI. 11.

and from A draw $AD \parallel$ to BC.   I. 31.

Then $\because AD$, BC are two $\parallel$ st. lines,

of which BC is $\perp$ to the given plane,

$\therefore AD$ is $\perp$ to the plane,   XI. 8.

and a line has been erected from $A \perp$ to the plane.

Q. E. F.

Proposition XIII. Theorem.

From the same point in a given plane, there cannot be two straight lines at right angles to the plane, upon the same side of it ; and there can be but one perpendicular to a plane from a point without the plane.

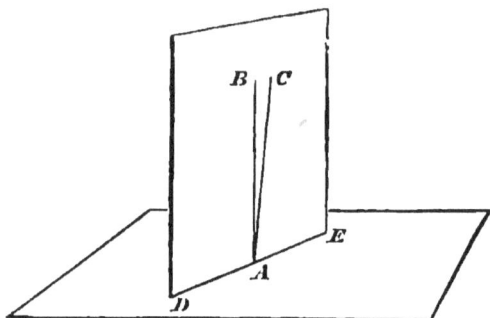

If it be possible, let two st. lines AB, AC, be at rt. ∠ s to a given plane, from the same pt. A in the plane, and upon the same side of it.

Let a plane pass through AB, AC: the common section of this with the given plane, is a st. line, passing through A.   XI. 2.

Let DAE be the common section of the planes.

Then the st. lines AB, AC, DAE are in one plane.

And ∵ CA is at rt. ∠ s to the given plane,

∴ CA is at rt. ∠ s to every st. line that meets it in that plane,   XI. Def. 2.

and DAE, which is in that plane, meets it ;

∠ CAE is a rt. ∠.

So also, ∠ BAE is a rt. ∠.

∴ ∠ CAE = ∠ BAE, in the same plane ; which is impossible.

Also, from a pt., without a plane, there can be but one perpendicular to that plane ; for if there could be two, they would be parallel to one another ; which is impossible.   XI. 6.

Planes, to which the same straight line is perpendicular, are parallel to one another.

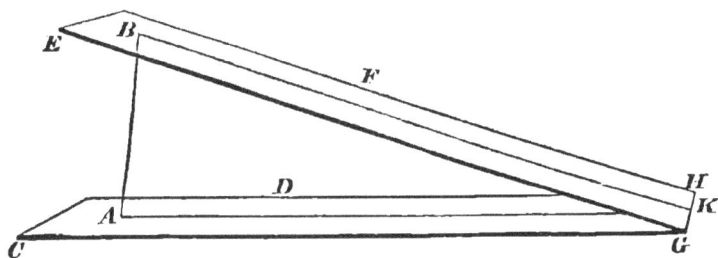

Let the st. line AB be ⊥ to each of the planes CD, EF.

Then must CD be parallel to EF.

If not, let them meet, and let the st. line GH be their common section.

In GH take any pt. K, and join AK, BK.

Then ∵ AB is ⊥ to the plane EF,

∴ AB is ⊥ to BK, a st. line in that plane, **XI. Def. 2.**

and ∴ ∠ ABK is a rt. ∠ .

So also, ∠ BAK is a rt. ∠ .

Hence two ∠ s of the △ ABK are together = two rt. ∠ s ; which is impossible. **I. 17.**

∴ the planes CD, EF do not meet when produced,

and ∴ CD is ∥ to EF. **XI. Def. 7.**

Q. E. D.

PROPOSITION XV. THEOREM.

If two straight lines, meeting one another, be parallel to two other straight lines, which meet one another, but are not in the same plane with the first two ; the plane, which passes through these, must be parallel to the plane passing through the others.

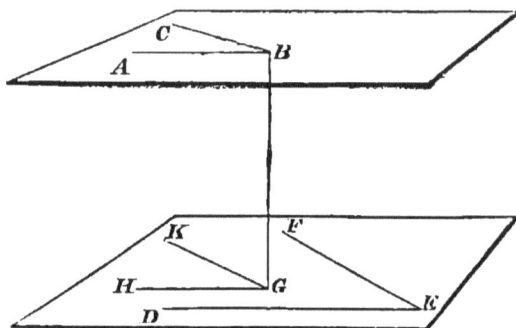

Let AB, BC, two st. lines meeting one another, be ‖ to DE, EF, which meet one another, but are not in the same plane with AB, BC.

Then must the plane AC be ‖ *to the plane DF.*

From B draw $BG \perp$ to the plane DF, meeting it in G. XI. 11.
Through G draw GH ‖ to ED, and GK ‖ to EF.   I. 31.
Then ∵ BG is $\perp$ to the plane DF,
 ∴ BG is $\perp$ to GH and GK, lines in that plane,
                       XI. Def. 2.
and ∴ each of the ∠ s BGH, BGK is a rt. ∠ .
Again ∵ BA and GH are both ‖ to ED,
 ∴ BA is ‖ to GH,   XI. 9.
and ∴ ∠ s GBA, BGH together = two rt. ∠ s.   I. 29.
 But ∠ BGH is a rt. ∠ .
 ∴ ∠ GBA is a rt. ∠ .
Hence GB is $\perp$ to BA ;
 and GB is $\perp$ to BC, for the same reason ;
 ∴ GB is $\perp$ to the plane AC.   XI. 4.
 Also, GB is $\perp$ to the plane DF ;   Constr.
 ∴ the plane AC is ‖ to the plane DF.   XI. 14.

                          Q. E. D.

If two parallel planes be cut by another plane, their common sections with it are parallel.

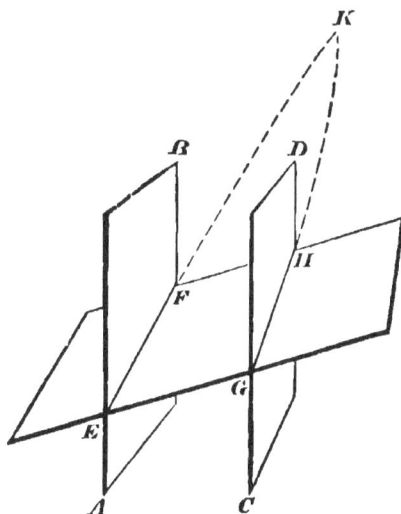

Let the parallel planes AB, CD be cut by the plane $EFHG$, and let their common sections with it be EF, GH.

Then must EF be $\parallel$ to GH.

If they be not $\parallel$, let them meet in K.

Then $\because EF$ is in the plane AB,

$\therefore K$ is a point in the plane AB.   XI. Def. 1.

So also, K is a point in the plane CD.   XI. Def. 1.

$\therefore$ the planes AB, CD meet, if produced.

But they do not meet, for they are parallel.

$\therefore EF$ and GH do not meet, when produced.

And EF, GH are in the same plane $EFGH$.

$\therefore EF$ is $\parallel$ to GH.   I. Def. 26.

Q. E. D.

PROPOSITION XVII.   THEOREM.

If two straight lines be cut by parallel planes, they must be cut in the same ratio.

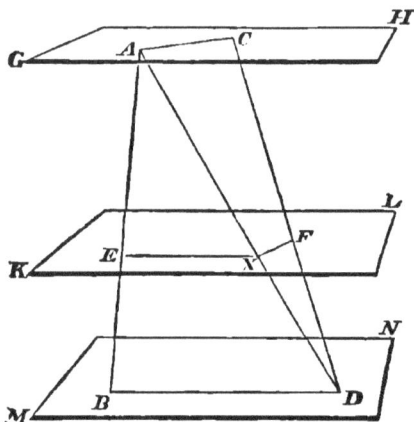

Let the st. lines AB, CD be cut by the ∥ planes GH, KL, MN in the pts. A, E, B; C, F, D.

Then must AE be to EB as CF is to FD.

Join AC, BD, AD.

Let AD meet the plane KL in the pt. X; and join EX, XF.

Then ∵ the ∥ planes KL, MN, are cut by the plane $EBDX$,

∴ EX is ∥ to BD.   XI. 16.

And ∵ the ∥ planes GH, KL, are cut by the plane $AXFC$,

∴ XF is ∥ to AC.   XI. 16.

Now ∵ EX is ∥ to BD, a side of △ ABD,

.. AE is to EB as AX is to XD ;   VI. 2.

and ∵ XF is ∥ to AC, a side of △ ADC,

∴ AX is to XD as CF is to FD.   VI. 2.

Hence AE is to EB as CF is to FD.   V. 5.

Q. E. D.

Proposition XVIII. Theorem.

If a straight line be at right angles to a plane, every plane, which passes through it, must be at right angles to that plane.

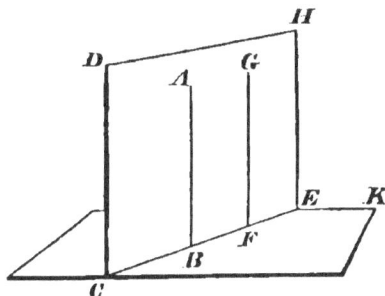

Let the st. line AB be ⊥ to the plane CK.

Then must every plane passing through AB be ⊥ to the plane CK.

Let any plane DE pass through AB, and let CE be the common section of the planes DE CK.

Take any pt. F in CE.

In the plane DE draw FG ⊥ to CE.   1. 11.

Then ∵ AB is ⊥ to the plane CK.

∴ AB is ⊥ to CE, a st. line in that plane ; XI. Def. 2.

and ∴ ∠ ABF is a rt. ∠ .

Now ∠ GFB is a rt. ∠ , by construction :

∴ FG is ∥ to AB.   1. 28.

And AB is ⊥ to the plane CK,

∴ FG is ⊥ to the plane CK.   XI. 8.

Then ∵ FG, a st. line in the plane DE, drawn ⊥ to CE, the common section of DE and CK, is ⊥ to CK,

∴ the plane DE is ⊥ to the plane CK.   XI. Def. 3.

So it may be proved that all planes, which pass through AB, are ⊥ to the plane CK.

Q. E. D.

PROPOSITION XIX. THEOREM.

If two planes, which cut one another, be each of them perpendicular to a third plane. their common section must be perpendicular to the same plane.

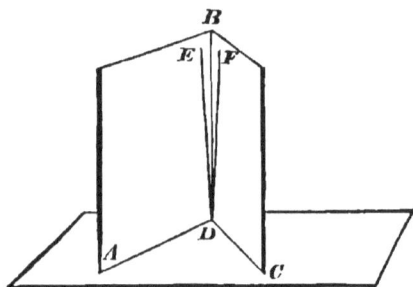

Let the two planes AB, BC be each ⊥ to a third plane, and let BD be the common section of AB and BC.

Then must BD be ⊥ to the third plane.

If it be not, draw, in the plane AB, the st. line DE ⊥ to AD, the common section of AB with the third plane ;   I. 11.

and draw, in the plane BC, the st. line DF ⊥ to DC, the common section of BC with the third plane.   I. 11.

Then ∵ the plane AB is ⊥ to the third plane,

and DE is drawn in the plane AB ⊥ to the common section,

∴ DE is ⊥ to the third plane.   XI. Def. 3.

So also, DF is ⊥ to the third plane.

Hence, from the pt. D, two st. lines are drawn ⊥ to the third plane, and on the same side of it; which is impossible. XI. 13.

∴ no other line but BD can be ⊥ to the third plane at D ;

∴ BD is ⊥ to the third plane.

Q. E. D.

PROPOSITION XX. THEOREM.

If a solid angle be contained by three plane angles, any two of them must be together greater than the third.

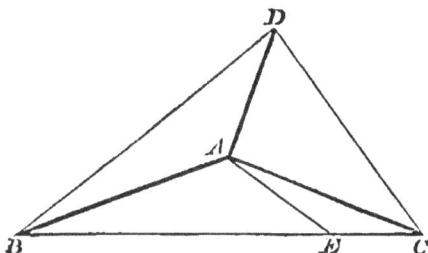

Let the solid ∠ at *A* be contained by the three plane ∠ s *BAC, CAD, DAB*.

Any two of these must be together greater than the third.

If the ∠ s *BAC, CAD, DAB*, be all equal, any two of them are together greater than the third.

If they are not equal, let *BAC* be that ∠, which is not less than either of the other two, and is greater than one of them, *DAB*.

At *A*, in the plane passing through *AB*, *AC*, make
∠ *BAE* = ∠ *DAB*,   I. 23.
    and make *AE* = *AD*, and through *E* draw the st. line *BEC*,
    cutting *AB*, *AC*, in the pts. *B*, *C* ; and join *DB*, *DC*.
Then in △ s *ABD, ABE*,
        ∵ *AD* = *AE*, and *AB* is common, and ∠ *BAD* = ∠ *BAE*,
        ∴ *DB* = *BE*.   I. 4.
Then ∵ *DB*, *DC* together are greater than *BC*,   I. 20.
    and *DB* = *BE*, a part of *BC*,
        ∴ *DC* is greater than *EC*.
Then in △ s *ADC, AEC*,
        ∵ *AD* = *AE*, and *AC* is common, and *DC* greater than *EC*,
        ∴ ∠ *DAC* is greater than ∠ *EAC*.   I. 25.
Also, by construction, ∠ *DAB* = ∠ *BAE*,
        ∴ ∠ s *DAC, DAB* together are greater than ∠ s *BAE*,
EAC together ;
    that is, ∠ s *DAC, DAB* together are greater than ∠ *BAC*.
Again, ∠ *BAC* is not less than either of the ∠ s *DAC, DAB*,
    and ∴ ∠ *BAC* with either of them is greater than the other.

Q. E. D.

Proposition XXI. Theorem.

Every solid angle is contained by plane angles, which are together less than four right angles.

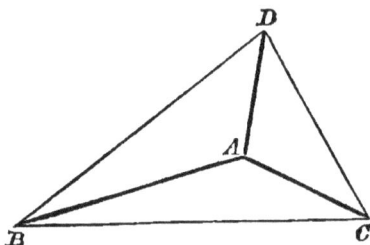

First, let the solid ∠ at *A* be contained by three plane ∠ s *BAC, CAD, DAB*.

These shall be together less than four right angles.

Take, in each of the st. lines *AB, AC, AD*, any points *B, C, D*, and join *BC, CD, DB*.

Then ∵ the solid ∠ at *B* is contained by the three plane ∠ s *CBA, ABD, DBC*,

∴ ∠ s *CBA, ABD* are together greater than ∠ *DBC*. XI. 20.

So also, ∠ s *BCA, ACD* are together greater than ∠ *BCD*,

and ∠ s *CDA, ADB* are together greater than ∠ *CDB*.

∴ the six ∠ s *CBA, ABD, BCA, ACD, CDA, ADB* are together greater than the three ∠ s *DBC, BCD, CDB*, and are ∴ together greater than two rt. ∠ s.

Again, ∵ the three ∠ s of each of the △ s *ABC, ACD, ADB* are together equal to two rt. ∠ s,   I. 32.

∴ the nine ∠ s *CBA, BAC, ACB, ACD, CDA, DAC, ADB, DBA, BAD* are together equal to six rt. ∠ s ; and of these the six ∠ s *CBA, ACB, ACD, CDA, ADB, DBA*, have been proved to be together greater than two rt. ∠ s,

and ∴ the three ∠ s *BAC, CAD, DAB*, which contain the solid ∠ at *A*, are together less than four rt. ∠ s.

Next, let the solid ∠ at *A* be contained by any number of plane ∠ s *BAC, CAD, DAE, EAF, FAB.*
These must be together less than four rt. ∠ s.

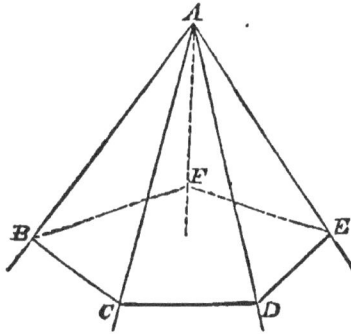

Let the planes, in which the ∠ s are, be cut by a plane, and let the common sections of it with those planes be *BC, CD, DE, EF, FB.*

Then ∵ the solid ∠ at *B* is contained by the three plane ∠ s *CBA, ABF, FBC*, of which any two are together greater than the third,   XI. 20.

∴ ∠ s *CBA, ABF* are together greater than ∠ *FBC.*

So also, the two plane ∠ s at each of the pts. *C, D, E, F,* which are at the bases of the △ s having the common vertex *A,* are together greater than the third ∠ at the same pt., which is one of the ∠ s of the polygon *BCDEF.*

∴ all the ∠ s at the bases of the △ s are together greater than all the ∠ s of the polygon.

Now all the ∠ s of the △ s together=twice as many rt. ∠ s as there are △ s, that is, as there are sides in the polygon *BCDEF* :   I. 32.

and all the ∠ s of the polygon, together with four rt. ∠ s, together=twice as many rt. ∠ s as there are sides in the polygon.   I. 32. Cor. 1

∴ all the ∠ s of the △ s together=all the ∠ s of the polygon together with four rt. ∠ s.

But all the ∠ s at the bases of the △ s have been proved to be together greater than all the ∠ s of the polygon ;

∴ all the ∠ s at the vertex *A* are together less than four rt. ∠ s.

Q. E. D.

Miscellaneous Exercises on Book XI.

1. If two straight lines in one plane, be equally inclined to another plane, they will be equally inclined to the common section of the two planes.

2. Two planes intersect at right angles in the line AB ; from a point C in this line are drawn CE and CF in one of the planes, so that the angle ACE is equal to BCF. Shew that CE and CF will make equal angles with any line through C in the other plane.

3. ABC is a triangle ; the perpendiculars from A, B, on the opposite sides, meet in D, and through D is drawn a straight line, perpendicular to the plane of the triangle ; if E be any point in this line, shew that EA, BC ; EB, CA ; and EC, AB ; are respectively perpendicular to each other.

4. A number of planes have a common line of intersection : what is the locus of the feet of perpendiculars on them from a given point ?

5. Two perpendiculars are let fall from any point on two given planes : shew that the angle between the perpendiculars will be equal to the angle of inclination of the planes to one another.

6. If perpendiculars AF, $A'F'$, be drawn to a plane from two points A, A', above it, and a plane be drawn through A perpendicular to AA', its line of intersection with the given plane is perpendicular to FF'.

7. Prove that equal straight lines drawn from a given point to a plane are equally inclined to the plane.

8. Prove that the inclination of a plane to a plane is equal to the angle between the perpendiculars to the two planes.

9. From a point above a plane two straight lines are drawn, the one at right angles to the plane, the other at right angles

to a given line in that plane : shew that the straight line joining the feet of the perpendiculars is at right angles to the given line.

10. In how many ways may a solid angle be formed with equilateral triangles and squares ?

11. Two planes are inclined to each other at a given angle. Cut them by a third plane, so that its intersections with the given planes shall be perpendicular to each other.

12. AB, AC, AD, are three given straight lines, at right angles to one another. AE is drawn perpendicular to CD, and BE is joined. Shew that BE is perpendicular to CD.

13. Two walls meet at any angle. Shew how to draw on their surfaces the shortest line joining a point on one to a point on the other.

14. Straight lines are drawn from two points to meet each other in a given plane. Find when their sum is the least possible.

15. If two parallel planes be cut by a third plane in the straight lines AB, ab, and by a fourth plane in the straight lines AC, ac respectively, the angle BAC will be equal to the angle bac.

16. If four points be so situated, that the distance between each pair is equal to the distance between the other pair, prove that the angles subtended at any one point by each pair of the others are together equal to two right angles.

17. Give a geometrical construction for drawing a straight line, which shall be equally inclined to three straight lines, meeting at a point.

18. A triangular pyramid stands on an equilateral base. The angles at its vertex are right angles. The square on the perpendicular from the vertex on the base is one-third of the square on either of the edges.

19. If one of the plane angles, forming a solid angle, be a right angle, and the sum of the other two be equal to two right angles, and a plane be drawn, cutting off equal lengths from the two edges, containing the right angle, the sum of the squares on the three straight lines, subtending the plane angles, will be double of the squares on the three edges, containing them.

20. If P be a point in a plane, which meets the containing edges of a solid angle in A, B, C, and O be the angular point, shew that the angles POA, POB, POC are together greater than half the angles AOB, BOC, COA, together.

BOOK XII.

If from the greater of two unequal magnitudes of the same kind there be taken more than its half, and from the remainder more than its half, and so on, there must at length remain a magnitude less than the smaller of the proposed magnitudes.

Let A and B be two unequal magnitudes of the same kind. of which A is the greater.

Then if from A there be taken more than its half, and from the remainder more than its half, and so on; there must at length remain a magnitude less than B.

Take a multiple of B, as mB, greater than A ; and divide A, by the process indicated, taking from it a magnitude greater than its half, and from the remainder a magnitude greater than its half, and carry this process on till there are m divisions, and call the parts successively taken away

C, D, E, F .... Z

Now $mB = B, B, B$ .... repeated m times,

and A is greater than the sum of $C, D, E,...Z......m$ in number.

Then Z, the last remainder, must be less than B.

For if not, since each of the preceding remainders is greater than Z, each of them would be greater than B, and the sum of $C, D.........Z$ would therefore be greater than mB ; that is, A would be greater than mB, which is contrary to the hypothesis.

∴ Z is less than B.

Q. E. D.

23

PROPOSITION I.   THEOREM.

Similar polygons inscribed in circles are to one another as the squares on the diameters of the circles.

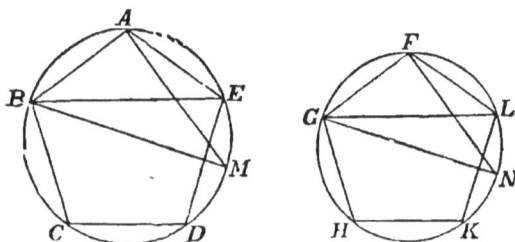

Let $ABCDE$, $FGHKL$ be similar polygons inscribed in two ⊙s, and let BM and GN be diameters of the ⊙s.

Then must polygon $ABCDE$ be to polygon $FGHKL$
        as sq. on BM is to sq. on GN.

Join AM, BE; FN, GL.

Then △ BAE is equiangular to △ GFL.   VI. 21.
∴ ∠ $AEB = $ ∠ FLG.

But ∠ $AMB = $ ∠ AEB, in the same segment,   III. 21.
and ∠ $FNG = $ ∠ FLG, in the same segment,
∴ ∠ $AMB = $ ∠ FNG.

also, ∠ $BAM = $ ∠ GFN, each being a rt. ∠ ,   III. 31.
∴ △ ABM is equiangular to △ FGN,
∴ AB is to BM as FG is to GN,   VI. 4.
and ∴ AB is to FG as BM is to GN.   V. 15.
∴ the duplicate ratio of AB to FG = the duplicate ratio
of BM to GN.   V. 21.

But polygon $ABCDE$ has to polygon $FGHKL$ the duplicate ratio of AB to FG.   VI. 21.

And sq. on BM has to sq. on GN the duplicate ratio of BM to GN.   VI. 21.

∴ polygon $ABCDE$ is to polygon $FGHKL$ as sq. on BM is to sq. on GN.   V. 5.

Q. E. D.

Circles are to one another as the squares on their diameters.

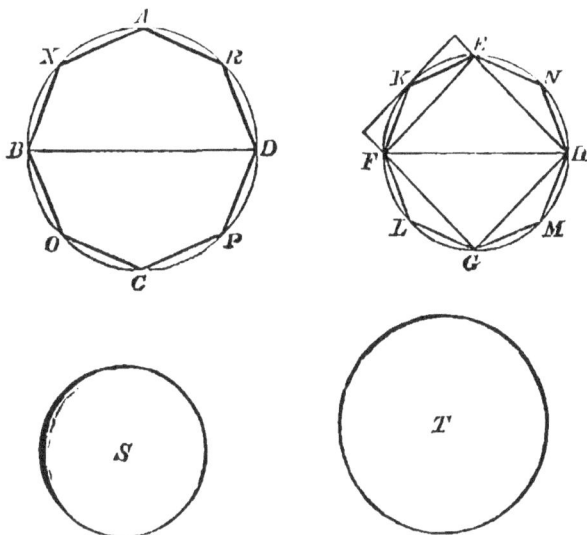

Let *ABCD, EFGH* be two ⊙s, and *BD, FH* their diameters :

Then must ⊙ *ABCD be to* ⊙ *EFGH as sq. on BD is to sq. on FH.*

For, if not, sq. on *BD* must be to sq. or *FH* as ⊙ *ABCD* is to some space either less than ⊙ *EFGH* or greater than it.

First, if possible, let it be as ⊙ *ABCD* is to a space *S* less than ⊙ *EFGH*.

In ⊙ *EFGH* describe the square *EFGH*.   IV. 6.

This square is greater than half of the ⊙ *EFGH*.

For the sq. *EFGH* is half of the square, which can be formed by drawing straight lines to touch the circle at the points *E, F, G, H* ; and the square thus formed is greater than the ⊙ :

∴ sq. *EFGH* is greater than half of the ⊙.

Bisect the arcs EF, FG, GH, HE at the pts. K, L, M, N, and join EK, KF, FL, LG, GM, MH, HN, NE.

Then each of the △s EKF, FLG, GMH, HNE, is greater than half of the segment of the circle in which it stands.

For △ EKF = half of the ☐, formed by drawing a st. line to touch the ⊙ at K, and parallel st. lines through E and F; and the ☐ thus formed is greater than the segment FEK;

∴ △ EKF is greater than half of the segment FEK, and similarly for the other △s.

∴ sum of all these triangles is greater than half of the sum of the segments of the ⊙, in which they stand.

Next, bisect EK, KF, etc., and form △s as before.

Then the sum of these △s is greater than half of the sum of the segments of the ⊙, in which they stand.

If this process be continued, and the △s be supposed to be taken away, there will at length remain segments of ⊙s, which are together less than the excess of the ⊙ $EFGH$ above the space S, by the Lemma.

Let segments EK, KF, FL, LG, GM, MH, HN, NE be those which remain, and which are together less than the excess of the ⊙ of the above S.

Then the rest of the ⊙, *i.e.* the polygon $EKFLGMHN$, is greater than S.

In ⊙ $ABCD$ inscribe the polygon $AXBOCPDR$ similar to the polygon $EKFLGMHN$.

The polygon $AXBOCPDR$ is to polygon $EKFLGMHN$ as sq. on BD is to sq. on FH,   XII. 1.

that is, as ⊙ $ABCD$ is to the space S.   Hyp. and V. 5.

But the polygon $AXBOCPDR$ is less than ⊙ $ABCD$,

∴ the polygon $EKFLGMHN$ is less than the space S;   V. 14. but it is also greater, which is impossible ;

∴ sq. on BD is not to sq. on FH as ⊙ $ABCD$ is to any space less than ⊙ $EFGH$.

In the same way it may be shown that

sq. on *FH* is not to sq. on *BD* as ⊙ *EFGH* is to any space less than ⊙ *ABCD*.

Nor is sq. on *BD* to sq. on *FH* as ⊙ *ABCD* is to any space greater than ⊙ *EFGH*.

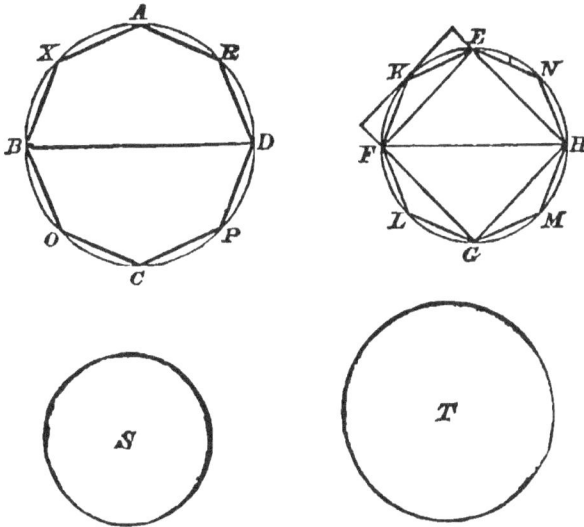

For, if possible, let it be as ⊙ *ABCD* is to a space *T*, greater than ⊙ *EFGH*.

Then, inversely, sq. on *FH* is to sq. on *BD* as space *T* is to ⊙ *ABCD*.

But as space *T* is to ⊙ *ABCD* so is ⊙ *EFGH* to some space, which must be less than ⊙ *ABCD*, because space *T* is greater than ⊙ *EFGH*.   V. 14.

∴ sq. on *FH* is to sq. on *BD* as ⊙ *EFGH* is to some space less than ⊙ *ABCD* ; which has been shewn to be impossible.

∴ sq. on *BD* is not to sq. on *FH* as ⊙ *ABCD* is to any space greater than ⊙ *EFGH*.

And it has been shown that

sq. on *BD* is not to sq. on *FH* as ⊙ *ABCD* is to any space less than ⊙ *EFGH*.

∴ sq. on *BD* is to sq. on *FH* as ⊙ *ABCD* is to ⊙ *EFGH*.

Q. E. D.

Papers on *Euclid (Books* VI., XI., *and* XII.) *set in the Cambridge Mathematical Tripos.*

1849.   VI. 4. Apply this proposition to prove that the rectangle, contained by the segments of any chord, passing through a given point within a circle, is constant.

XI. 11. Prove that equal right lines, drawn from a given point to a given plane, are equally inclined to the plane.

1850.   VI. 10. AB is a diameter, and P any point in the circumference of a circle ; AP and BP are joined and produced, if necessary ; if from any point C of AB a perpendicular be drawn to AB, meeting AP and BP in points D and E respectively, and the circumference of the circle in a point F, shew that CD is a third proportional to CE and CF.

1851.   VI. 3. If A, B, C be three points in a straight line, and D a point, at which AB and BC subtend equal angles, show that the locus of the point D is a circle.

XI. 8. From a point E draw EC, ED perpendicular to two planes CAB, DAB, which intersect in AB, and from D draw DF perpendicular to the plane CAB, meeting it in F: shew that the line, joining the points C and F, produced if necessary, is perpendicular to AB.

1352.   VI. 2. If two triangles be on equal bases, and between the same parallels, any line, parallel to their bases, will cut off equal areas from the two triangles.

1852.   XI. 11. $ABCD$ is a regular tetrahedron, and, from
the vertex A, a perpendicular is drawn to
the base BCD, meeting it in O: shew that
three times the square on AO is equal to twice
the square on AB.

1853.   VI. 6.   If the vertical angle C, of a triangle ABC, be
bisected by a line, which meets the base in
D, and is produced to a point E, such that
the rectangle, contained by CD and CE, is
equal to the rectangle, contained by AC and
CB: shew that if the base and vertical angle
be given, the position of E is invariable.

XI. 21.   If BCD be the common base of two pyramids,
whose vertices A and A' lie in a plane passing through BC, and if the two lines AB, AC,
be respectively perpendicular to the faces
$BA'D$, $CA'D$, prove that one of the angles at
A, together with the angles at A', make up
four right angles.

1854.   VI. 16. EA, EA' are diameters of two circles, touching
each other externally at E; a chord AB of
the former circle, when produced, touches the
latter at C', while a chord $A'B'$ of the latter
touches the former at C: prove that the rectangle, contained by AB and $A'B'$, is four
times as great as that contained by BC' and
$B'C$.

XI. 20.   Within the area of a given triangle is described
a triangle, the sides of which are parallel to
those of the given one: prove that the sum
of the angles, subtended by the sides of the
interior triangle, at any point, not in the plane
of the triangles, is less than the sum of the
angles, subtended at the same point by the
sides of the exterior triangle.

1855.   VI. 2.   A tangent to a circle, at the point A, intersects
two parallel tangents in B, C, the points of

contact of which with the circle are D, E, respectively: shew that if BE, CD, intersect in F, AF is parallel to the tangents BD, CE.

1855.   XI. 16. From the extremities of the two parallel straight lines AB, CD, parallel lines Aa, Bb, Cc, Dd, are drawn, meeting a plane in a, b, c, d : prove that AB is to CD as ab is to cd, taking the case, in which A, B, C, D are on the same side of the plane.

1856.   VI. Def. 1. Enunciate the propositions, which prove that in the case of triangles the conditions of similarity are not independent.

XI. 11. Shew that the perpendicular, dropped from the vertex of a regular tetrahedron upon the opposite base, is treble of that dropped from its own foot upon any of the other bases.

1857.   VI. 19. Any two straight lines, BB', CC', drawn parallel to the base DD', of a triangle ADD', cut AD in B, C, and AD' in B', C' ; BC', $B'C$, are joined. prove that the area ABC' or $AB'C$ varies as the rectangle, contained by BB', CC'.

XI. 16. A triangular pyramid stands on an equilateral base, and the angles at the vertex are right angles : shew that the sum of the perpendiculars on the faces, from any point of the base, is constant.

1858.   VI. 15. Find a point in the side of a triangle, from which two lines, drawn one to the opposite angle, and the other parallel to the base, shall cut off, towards the vertex and towards the base, equal triangles.

XI. 11. Two planes intersect : shew that the loci of the points, from which perpendiculars on the planes are equal to a given straight line, are straight lines ; and that four planes may be

drawn, each passing through two of these
lines, such that the perpendiculars, from any
point in the line of intersection of the given
planes, upon any one of the four planes, shall
be equal to the given line.

1859. **vi. 31.** Shew that, on a given straight line, there may
be described as many polygons of different
magnitudes, similar to a given polygon, as
there are sides of different lengths in the
polygon.

xi. 20. Three straight lines, not in the same plane,
intersect in a point, and through their point
of intersection another straight line is drawn
within the solid angle formed by them : prove
that the angles, which this straight line makes
with the first three, are together less than the
sum, but greater than half the sum of the
angles which the first three make with each
other.

1860. **vi. A.** If the two sides, containing the angle, through
which the bisecting line is drawn, be equal,
interpret the result of the proposition.
Prove from this proposition and the preceding,
that the straight lines, bisecting one angle of
a triangle internally, and the other two externally, pass through the same point.

xi. 17. If three straight lines, which do not all lie in
one plane, be cut in the same ratio by three
planes, two of which are parallel, shew that
the third will be parallel to the other two, if
its intersections with the three straight lines
are not all in one straight line.

1861. **vi. 6.** From the angular points of a parallelogram
ABCD, perpendiculars are drawn on the
diagonals, meeting them in *E, F, G, H* re-

spectively ; prove that $EFGH$ is a parallelogram similar to $ABCD$.

1861.   XI. 12. Shew that the shortest distance between two
opposite edges of a regular tetrahedron is
equal to half the diagonal of the square, described on an edge.

1862.   VI. 1. Lines are drawn from two of the angular points
of a triangle, to divide the opposite sides in
a given ratio ; prove that the line, joining
the third angular point with the point of intersection of these two lines, either bisects
the opposite side, or divides it in a ratio
which is the duplicate of the given ratio.

XI. 21. If four points be so situated that the distance
between each pair is equal to the distance
between the other pair, prove that the angles
subtended at any one of these points by each
pair of the others, are together equal to two
right angles.

1863.   VI. 4. The internal angles at the base of a triangle, and
the external angle at the vertex, are bisected
by straight lines ; prove that the three points,
in which these straight lines meet the opposite sides respectively, lie on one straight
line.

XI. 17. If each edge of a tetrahedron be equal to the
opposite edge, the straight line, joining the
middle points of any two opposite edges,
shall be at right angles to each of those
edges.

1864.   VI. 23. If one parallelogram have to another parallelogram the ratio, which is compounded of the
ratios of their sides, the parallelograms shall
be equiangular.

XI. 12. On a given equilateral triangle describe a
regular tetrahedron.

1865. **VI. 19.** The opposite sides, BA, CD of a quadrilateral $ABCD$, which can be inscribed in a circle, meet, when produced, in E ; F is the point of intersection of the diagonals, and EF meets AD in G: prove that the rectangle EA, AB is to the rectangle ED, DC as AG is to GD.

XI. 16. In the triangular pyramid $ABCD$, AB is at right angles to CD, and AC to BD: prove that AD is at right angles to BC.

1866. **VI. 4.** ABC is an isosceles triangle ; AE is the perpendicular from A on the base BC; D is any point in AE ; and CD produced meets the side AB at F: shew that the ratio of AD to DE is double of the ratio of AF to FB.

XII. 1. Give an outline of Euclid's demonstration that circles are to one another as the squares on their diameters.

1867. **VI. A.** Each acute angle of a right-angled triangle and its corresponding exterior angle are bisected by straight lines meeting the opposite sides ; prove that the rectangle, contained by the portions of those sides intercepted between the bisecting lines is four times the square on the hypotenuse.

XI. 21. Two pyramids are described, the one standing on a square as a base, the other on a regular octagon, the vertex of each being equally distant from the angular points of its base ; if this distance be the same for each pyramid, and the perimeters of the bases be equal, prove that the plane angles, containing the solid angle at the vertex of the former, are together greater than the plane angles, containing the solid angle at the vertex of the latter.

1868. **VI. 2.** Without assuming any subsequent proposition, prove that the equiangular triangles in either

of the figures of this proposition, are to each
other in the duplicate ratio of the sides opposite to the equal angles.

1868. XI. 11. Of the least angles, which a given line in one
plane makes with any line in another plane,
the greatest for different positions of the
given line is that which measures the inclination of the two planes.

1869. XI. 20. If O be a point, within a tetrahedron $ABCD$,
prove that the three angles of the solid angle,
subtended by BCD at O, are together greater
than the three angles of the solid angle at A.

1870. VI. 15. Two straight lines are given in position, and a
third straight line is drawn so as to cut oft
a triangle equal to a given triangle ; through
the middle point of this third side is drawn
a straight line in a given direction, terminated by the two given straight lines: prove
that the rectangle under the segments of the
intercepted part is constant.

XI. 7. In a tetrahedron each edge is perpendicular to
the direction of the opposite edge ; prove
that the straight line joining the centre of
the sphere, circumscribing the tetrahedron,
to the middle point of any edge, is equal and
parallel to the straight line joining the centre
of perpendiculars to the middle point of the
opposite edge.

1371. VI. 2. ABC is a triangle, and lines AO, BO, CO cut
the opposite sides in D, E, F ; if EF cut BC
in G, prove that BD is to DC as BG is to
GC.

XI. 11. The perpendiculars from the angular points of
a tetrahedron on the opposite faces meet in a
point : prove that the necessary and sufficient
condition for this is that the sums of the
squares on pairs on opposite edges be equal.

1872. VL 2. Draw through a point a straight line, so that the part of it intercepted between a given straight line and a given circle may be divided at the given point in a given ratio. Between what limits must the ratio lie in order that a solution may be possible?

XL 20. If the opposite edges of a tetrahedron be equal two and two, prove that the faces are acute-angled triangles. Prove also that a tetrahedron can be formed of any four equal and similar acute-angled triangles.

APPENDIX.

EXAMINATION PAPERS IN EUCLID

SET TO CANDIDATES FOR

First and Second Class Provincial Certificates,

AND TO STUDENTS MATRICULATING IN THE

UNIVERSITY OF TORONTO.

SECOND CLASS PROVINCIAL CERTIFICATES, 1871.
TIME—TWO HOURS AND A HALF.

1. If two triangles have two sides of the one equal to two sides of the other, each to each, and have likewise their bases equal, the angle which is contained by the two sides of the one shall be equal to the angle contained by the two sides, equal to them, of the other.

2. Triangles upon the same base, and between the same parallels, are equal to one another.

3. If the square described upon one of the sides of a triangle be equal to the squares described upon the other two sides of it, the angle contained by these two sides is a right angle.

4. If a straight line be divided into two equal, and also into two unequal, parts, the squares on the two unequal parts are together double of the square on half the line, and of the square on the line between the points of section.

6. If a straight line be divided into any two parts, the rectangles contained by the whole and each of the parts are together equal to the square on the whole line.

6. Bisect a parallelogram by a straight line drawn from a point in one of its sides.

7. Let A B C be a triangle, and let D D be a straight line drawn to D, a point in A C between A and C, then, if A B be greater than A C, the excess of A B above A C is less than that of B D above D C.

8. In a triangle A B C, A D being drawn perpendicular to the straight line B D which bisects the angle B, show that a line drawn from D parallel to B C will bisect A C.

NOTE.—The percentage of marks requisite, in order that a candidate may be ranked of a particular grade, will be taken on the value of the above paper, omitting question 8.

ii.	APPENDIX.

SECOND CLASS PROVINCIAL CERTIFICATES, 1872.

TIME—2¾ HOURS.

1. Define a *straight line*, a *plane rectilineal angle*, a *right angle*, a *Gnomon*. Enunciate Euclid's Postulates.
2. If from the ends of the side of a triangle there be drawn two straight lines to a point within the triangle, these shall be less than the other two sides of the triangle, but shall contain a greater angle.
3. If two triangles have two angles of the one equal to angles of the other, each to each, and one side equal to one side, namely, either the sides adjacent to the equal angles, or sides which are opposite to equal angles in each; then shall the other sides be equal, each to each; and also the third angle of the one equal to the third angle of the other. (*Take the case in which the assumed equal sides are those opposite to equal angles.*)
4. In every triangle, the square on the side subtending an acute angle is less than the sides containing that angle, by twice the rectangle contained by either of these sides, and the straight line intercepted between the perpendicular let fall on it from the opposite angle, and acute angle. (*Take the case where the perpendicular falls within the triangle.*)
5. If a straight line be divided into any two parts, the squares on the whole line, and one of the parts, are equal to twice the rectangle contained by the whole and that part, together with the square on the other part.
6. Prove that, if a straight line AD be drawn from A, one of the angles of a triangle ABC, to D, the middle point of the opposite side BC, BA × AC is greater than 2 AD.
7. Let the equilateral triangle ABC, and triangle ADB, in which the angle ABD is a right angle, be on the same base AB, and between the same parallels AB and CD. Prove that $4 AD^2 = 7 AB^2$
8. From D, a point in AB, a side of the triangle ABC, it is required to draw a straight line DE, cutting BC in E, and AC produced in F, so that DE may be equal to EF.

SECOND CLASS PROVINCIAL CERTIFICATES, 1873.

TIME—TWO HOURS AND A HALF.

NOTE.—Candidates who take only Book I, will confine themselves to the first eight questions; those who take Books I and II, will omit the first two questions.

1   If two angles of a triangle be equal to one another, the
     sides also which subtend, or are opposite to, the equal
     angles, shall be equal to one another.
2.   If one side of a triangle be produced, the exterior angle
     shall be greater than either of the interior opposite
     angles.
3.   The opposite side, and angles of a parallelogram, are
     equal to one another.
4.   The complements of the parallelograms, which are about
     the diameter of any parallelogram, are equal to one
     another.
5.   To describe a square on a given straight line.
6.   Let A B C D be a quadrilateral figure whose opposite
     angles A B C and A D C are right angles.   Prove
     that, if A B be equal to A D, C B and C D shall also
     be equal to one another.
7   If A B C D be a quadrilateral figure, having the side A B
     parallel to the side C D, the straight line which joins
     the middle points of A B and D C shall divide the
     quadrilateral into two equal parts.
8.   The straight line, which joins the middle points of two
     sides of a triangle, is parallel to the base.
9.   If a straight line be divided into any two parts, the
     square on the whole line is equal to the squares on the
     two parts, together with twice the rectangle contained
     by the parts.
10. In an obtuse angled triangle, is the sum of the sides con-
     taining the obtuse angle greater or less than the
     square of the side opposite to the obtuse angle?
     And, by how much?   Prove the proposition.

SECOND CLASS PROVINCIAL CERTIFICATES, 1874.

TIME— TWO HOURS AND THREE-QUARTERS.

NOTE.—Candidates who take only Book I. will confine themselves to the first 7 questions.   Those who take Books I. and
II. will omit questions 1, 2, and 3.

1.   When is one straight line said to be *perpendicular to an-*
     other.
     To draw a straight line perpendicular to a given
     straight line of an unlimited length, from a given
     point without it.
2.   If one side of a triangle be produced, the exterior angle
     shall be greater than either of the interior opposite
     angles.
     two triangles have two angles of the one equal to two
     angles of the other, each to each; and one side
     equal to one side, namely, sides which are opposite to

equal angles in each; then shall the other
equal, each to each.

4. What are *parallel straight lines ?*
   If a straight line, falling on two other straight lines,
   make the alternate angles equal to one another, the
   two straight lines shall be parallel to one another.

5. What is a *parallelogram ?*
   Parallelograms on equal bases, and between the same
   parallels, are equal to one another.

6. If two isosceles triangles be on the same base, and on the
   same side of it, the straight line which joins their
   vertices, will, if produced, cut the base at right angles.

7. Let ABC be a triangle, in which the angle ABC is a right
   angle. From AC cut off AD equal to AB, and join
   BD. Prove that the angle BAC is equal to twice the
   angle CBD.

8. If a straight line be divided into two equal parts and also
   into two unequal parts, the rectangle contained by
   the unequal parts, together with the square on the line
   between the points of section, is equal to, &c. (5, II.)

9. In every triangle, the square on the side subtending an
   acute angle is less than the squares on the sides containing that angle, by &c. (13, II). (It will be suf-
   ficent to take the case in which the perpendicular falls
   within the triangle.)

10. To describe a square that shall be equal to a given recti-
    lineal figure.

11. The square on any straight line drawn from the vertex of
    an isosceles triangle to the base is less than the
    square on a side of a triangle by a rectangle contained
    by the segments of the base.

SECOND CLASS PROVINCIAL CERTIFICATES, 1875.

TIME—TWO HOURS AND THREE-QUARTERS.

NOTE.—Those students who take only Book I. will confine
themselves to the first seven questions. Those who take
Books I. and II. will omit the questions marked with an
asterisk (*), namely, (1) and (2).

*1. If one side of a triangle be produced, the exterior angle
    is greater than either of the interior opposite angles.

*2. If two triangles have two angles of the one equal to two
    angles of the other, each to each, and one side equal
    to one side, namely, the sides opposite to equal angles,
    then shall the other sides be equal, each to each.

3. If a straight line falling on two other straight lines make
   the alternate angles equal to each other, these two
   straight lines shall be parallel.

4. If a straight line fall upon two parallel straight lines, it makes the two interior angles upon the same side together equal to two right angles.
5. Assuming Proposition XXXII, deduce the corollary: "all the exterior angles of any rectilineal figure, made by producing the sides successively in the same direction, are together equal to four right angles."
6. If a straight line, drawn parallel to the base of a triangle, bisect one of the sides, it shall bisect the other also.
7. Let ABC and ADC be two triangles on the same base AC and between the same parallels AC and BD. Prove, that, if the sides AB and BC be equal to one another, their sum is less than the sum of the sides AD and DC.
8. If a straight line be divided into any two parts, the rectangles contained by the whole and each of the parts are together equal to the square on the whole line.
9. If a straight line be bisected and produced to any point, the rectangles contained by the whole line thus produced, and the part of it produced, together with, etc., (6, II).
10. Divide a straight line into two parts, such that the sum of their squares may be the least possible.

FIRST CLASS PROVINCIAL CERTIFICATES, 1871.

1. To describe a square that shall be equal to a given rectilineal figure.
2. A segment of a circle being given, to describe the circle of which it is the segment.
3. If the vertical angle of a triangle be divided into two equal angles by a straight line which also cuts the base, the segments of the base shall have the same ratio which the other sides of the triangle have to one another.
4. In a right-angled triangle, if a perpendicular be drawn from the right angle to the base, the triangles on each side of it are similar to the whole triangle and to one another.
5. If four straight lines be proportionals, the similar rectilineal figures similarly described upon them shall also be proportionals.
6. Draw a straight line so as to touch two given circles.
7. Let A B C be a triangle, and from B and C, the extremities of the base B C, let line B F and C E be drawn to F and E, the middle points of A C and A B respect

ively, then, if B F = C E, A B and A C shall be equal
to one another.

8. Describe an equilateral triangle equal to a given triangle.

FIRST CLASS PROVINCIAL CERTIFICATES, 1872.

TIME—TWO AND A HALF HOURS.

1. If a straight line touch a circle, and from the point of contact a straight line be drawn cutting the circle, the angles which this line makes with the line touching the circle shall be equal to the angles which are in the alternate segments of the circle.

2. To inscribe a circle in a given triangle.

3. Equal triangles which have one angle of the one equal to one angle of the other, have their sides about the equal angles reciprocally proportional.

4. Similar triangles are to one another in the duplicate ratio of their homologous sides.

5. In any right angled triangle, any rectilineal figure described on the side subtending the right angle is equal to the similar and similarly described figures on the sides containing the right angle.

6. Two circles cut each other, and through the points of section are drawn two parallel lines, terminated by the circumferences. Prove that these lines are equal.

7. Let A C and B D, the diagonals of a quadrilateral figure A B C D, intersect in E. Then, if A B be parallel to C D, the circles described about the triangles A B E and C D E shall touch one another.

8. Divide a triangle into two equal parts by a straight line at right angles to one of the sides.

FIRST CLASS PROVINCIAL CERTIFICATES, 1873.

TIME—THREE HOURS.

1. The angle in a semicircle is a right angle.

2. A segment of a circle being given, describe the circle of which it is a segment.

3. Give Euclid's definition of proportion; and prove, by taking equi-multiples according to the definition, that 2, 3, 9, 13, are not proportionals.

4. Similar triangles are to one another in the duplicate ratio of their homologous sides.

5. To find a mean proportional between two given straight lines.

6. Through C, the vertex of a triangle A C B, which has the sides A C and C B equal to one another, a line C D

is drawn parallel to A B; and straight lines, A D,
D B, are drawn from A and B to any point D in C D.
Prove that the angle A C D is greater than the angle
A D B.

7 A B C D is a quadrilateral figure inscribed in a circle.
From A and B, perpendiculars A E, B F are let fall
on C D (produced if necessary); and from C and D,
perpendiculars C G, D H, are let fall on B A (produced
if necessary). Prove that the rectangles A E, B F and
C G, D H, are equal to one another.

8 A B C D is a quadrilateral figure inscribed in a circle. The
straight line D E drawn through D parallel to A B,
cuts the side B C in E ; and the straight line A E produced meets D C produced in F. Prove, that if the
rectangle B A, A D be equal to the rectangle E C, C F,
the triangle A D F shall be equal to the quadrilateral
A B C D.

FIRST CLASS PROVINCIAL CERTIFICATES, 1874.

TIME—THREE HOURS.

1. In equal circles, equal straight lines cut off equal circumferences, the greater, equal to the greater, and the less
to the less.

2. To describe a circle about a given equilateral and equiangular pentagon.

3. To find a mean proportional between two given straight
lines.

4. What is meant by duplicate ratio ? Write down two whole
numbers, which are in the duplicate ratio of ½ to ⅓.
What are similar rectilineal figures ?
Similar triangles are to one another in the duplicate ratio
of their homologous sides.

5. In any right angled triangle, any rectilineal figure described
on the side subtending the right angle is equal to the
similar and similarly described figures on the sides
containing the right angle.

6 To describe a triangle, of which the base, the vertical angle,
and the sum of the two sides are given.

7. From A the vertex of a triangle ABC, in which each of the
angles ABC and ACB is less than right angle, AD is
let fall perpendicular on the base BC. Produce BC to
E, making CE equal to AD ; and let F be a point in
AC, such that the triangle BFE is equal to the triangle ABC. Prove that F is one of the angular
points of a square inscribed in the triangle ABC,
with one of its sides on BC.

8. Let E be the point of intersection of the diagonals of a
quadrilateral figure ABCD, of which any two opposite
angles are together equal to two right angles. Produce BC to G, making CG equal to EA; and produce
AD to F, making DF equal to BE. Prove that if EG
and EF be joined, the triangles EDF and ECG are
equal to one another.

FIRST CLASS PROVINCIAL CERTIFICATES, 1875.

TIME—THREE HOURS.

1. If two triangles have two angles of the one equal to two
angles of the other, each to each, and one side equal
to one side, namely, the sides adjacent to the equal
angles in each, then shall the other sides be equal
each to each.
2. From a given circle to cut off a segment, which shall contain an angle equal to a given rectilineal angle.
3. If the angle of a triangle be divided into two equal angles
by a straight line which also cuts the base, the
segments of the base shall have the same ratio which
the other sides of the triangles have to one another.
4. The sides about the equal angles equi-angular triangles are
proportionals; and those which are opposite to the
equal angles are homologous sides.
5. If the similar rectilineal figures similarly described upon
four straight lines be proportionals, those straight
lines shall be proportionals.
6. Any rectangle is half the rectangle contained by the
diameters of the squares on its adjacent sides.
7. Through a given point within a given circle, to draw a
straight line such that one of the parts of it intercepted between that point and the circumference shall be
double of the other.
8. If, from any point in a circular arc, perpendiculars be let
fall on its bounding radii, the distance of their feet is
invariable.

MATRICULATION, 1871.

1. State the points of agreement and disagreement of the
circle, square and rhombus, with one another as
appearing from their definitions.
2. Any two sides of a triangle are together greater than the
third side.
Show that the sum of the excesses of each pair of sides
above the third side is equal to the sum of the three
sides of the triangle.

2. If the square described upon one of the sides of a triangle be equal to the square described on the other two sides of it, the angle contained by these two sides is a right angle.

In an isosceles triangle if the square on the base be equal to three times the square on either side the vertical angle is two-thirds of two right angles.

4. If a straight line be divided into any two parts the square on the whole line is equal to the square on the two parts, together with twice the rectangle contained by the parts.

Is there any difference between the principle of this proposition and the statement $(a + b)^2 = a^2 + 2ab + b^2$.

Of all the squares that can be inscribed within another the least is that formed by joining the bisections of the side.

5. If a straight line be divided into two equal and also into two unequal parts, the squares on the two unequal parts are together double of the square on half the line and of the square on the line between the points of section.

Does the statement respecting the equality of the square hold for any other division of the line.

6. Equal straight lines in a circle are equally distant from the centre; and conversely, those which are equally distant from the centre are equal to one another.

The lines joining the extremities of two equal straight lines in a circle towards the same parts are parallel to each other.

7. What is meant by the Angle in a segment of a circle? Define similar segments of circles.

Upon the same straight line and upon the same side of it, there cannot be two similar segments of circles not coinciding with one another.

8. In equal circles the angles which stand upon equal arcs, are equal to one another whether they be at the centres or circumferences.

If two equal circles so intersect each other that the tangents at one of their points of intersection are inclined to each other at an angle of 60° shew that

Radius of circle : line joining their centres : : $1 : \sqrt{3}$.

9. From a given circle to cut off a segment that shall contain an angle equal to a given rectilineal angle.

In a given circle inscribe a triangle which shall have a given vertical angle, and whose area shall be equal to a given triangle; and shew with what limitation this can be done.

10. When is a circle said to be inscribed in a rectilinea
     figure.
     To inscribe a circle in a given triangle.
11. Inscribe an equilateral and equiangular pentagon in a
     given circle.
     Show how to divide a right angle into fifteen equal parts

---

MATRICULATION, 1872.

HONORS.

1. From a given point to draw a straight line equal to a given
   straight line.
   Explain what different constructions there are in this
   proposition.
2. If a side of a triangle be produced, the exterior angle is
   equal to the two interior and opposite angles; and the
   three interior angles of every triangle are together
   equal to two right angles.
   Find the number of degrees in one of the exterior angles
   of a regular heptagon.
3. Triangles upon the same or equal bases and between the
   same parallels are equal to one another.
   By means of these propositions prove that a line drawn
   parallel to the base of a triangle and cutting off one-
   fourth from one of its sides, will also cut off a fourth
   part from the other side.
4. If a straight line be divided into two equal and also into
   two unequal parts, the squares on the two unequal
   parts are together double of the square on half the
   line, and of the square on the line between the points
   of section.
   If a chord be drawn parallel to the diameter of a circle
   and from any point in the diameter lines be drawn to
   its extremities, the sum of their squares will be equal
   to the sum of the squares of the segments of the diameter.
5. To divide a given straight line into two parts, so that the
   rectangle contained by the whole and one of the parts
   shall be equal to the square on the other part.
   Solve the problem algebraically. Interpret and construct
   geometrically the second root so obtained.
   Divide a given line so that one segment may be a geometric mean between the whole and the other.
6. In every triangle, the square on the side subtending either
   of the acute angles, is less than the squares on the
   sides containing that angle, by twice the rectangle
   contained by either of these sides, and the straight

line intercepted between the acute angle and the perpendicular let fall upon it from the opposite angle.

In a triangle ABC, if AD be drawn to the bisection of BC, the difference between the square on BC and twice the square on AC is double of the difference between the square on AB, and twice the square on AD.

7. If a straight line touch a circle, the straight line drawn from the centre to the point of contact shall be perpendicular to the line touching the circle.

The locus of intersections of all pairs of tangents to a circle which contain a given angle is a circle.

What is the magnitude of this angle, in order that the circle may be double the original ?

8. The opposite angles of any quadrilateral figure inscribed in a circle are together equal to two right angles.

What relation must exist between the sides of a quadrilateral in order that a circle may be inscribed in it ? Show that your relation is sufficient.

9. If from any point without a circle two straight lines be drawn, one of which cuts the circle, and the other touches it ; the rectangle contained by the whole line which cuts the circle, and the part of it without the circle, shall be equal to the square on the line which touches it.

Show that this proposition is an extension of III, 36.

From a given point without a circle show how to draw (when possible) a line that will be divided by that circle in Medial section.

10. Inscribe a circle in a given triangle.

When is one rectilineal figure said to be inscribed in another.

11. In a right-angled triangle, if the perpendicular be drawn from the right angle to the base ; the triangle on each side of it are similar to the whole triangle and to one another.

Construct geometrically the roots of the equation $x(a-x)=b^2$ and give the geometric interpretation of the case of equal and impossible roots that the problem may present.

12. To describe a rectilineal figure which shall be similar to one given rectilineal figure and equal to another given rectilineal figure.

MATRICULATION, 1873.

HONORS.

1. If a straight line falls upon two parallel straight lines, it makes the alternate angles equal to one another, and

the exterior angle equal to the interior and opposite upon the same side, and also the two interior angles upon the same side together equal to two right angles. Vary the order of proof in this proposition by proving the last statement first.

2. If a straight line falling upon two other straight lines, makes the interior angles upon the same side together equal to two right angles, the two straight lines shall be parallel to one another.
Can this be inferred immediately from the 12th axiom? Give the reasons for your answer.

3. Any two sides of a triangle are together greater than the third side.
A straight line is the shortest distance between two given points.

4. In any right angled triangle, the square which is described upon the side subtending the right angle, is equal to the squares described upon the sides which contain the right angle.
Any two parallelograms being described on two sides of any triangle, to describe on the third side a parallelogram equal to their sum.

5. To describe a square that shall equal a given rectilineal figure.
To divide a given straight line into two parts such that their rectangle is equal to a given rectilineal figure.
What limitation must there be to the magnitude of the given figure?

6. If a straight line drawn through the centre of a circle bisect a straight line in it which does not pass through the centre, it shall cut it at right angles; and, if it cut it at right angles, it shall bisect it.
Describe three circles of given radii which shall touch each other externally two and two.

7 In the above show that the common tangents meet in one point, with which as centre, a circle may be described passing through the three points of contact.
What proposition of Euclid does this correspond to?

8. If straight lines within a circle intersect in one point the rectangle under the segments is constant.
What limitation must be made to render the converse true? Prove the converse when true.

9. The opposite angles of any quadrilateral figure inscribed in a circle are together equal to two right angles.
Deduce—The angle in a semicircle is a right angle. (Prop. 31 Bk. III.)

10. To describe an isosceles triangle having each of the angles at the base double of the third angle.

A tangent to a circle is drawn at an angular point of an inscribed regular pentagon, and a side produced through that point, show that a straight line making equal intercepts on the tangent and the side produced, is parallel to the tangent at one of the adjacent angular points.

11. To describe a circle about a given equilateral pentagon. With an angular point of the regular pentagon as centre, and a side as radius, describe a second circle; show that the tangent to the first circle at a point of intersection of the circles meets the common diameter at a point without the second circle.

12. In the above show that the distance from the above point to the centre of the first circle is greater than the diameter of the second circle.

MATRICULATION, 1874.

HONORS.

₊ Nos. 1 and 3 to be omitted for Senior Matriculation ; Nos. 12 and 13 to be omitted for Junior Matriculation.

1. Parallelograms upon the same base and between the same parallels are equal to one another.
From the centre O of a circle the radii OA, OB are drawn, the tangents at A and B meet in C; if OC be bisected in D and DE be drawn perpendicular to OD meeting OB in E, then AE will bisect the figure $OBCA$.

2. In every triangle the square on the side subtending any of the acute angles is less than the squares on the sides containing that angle by twice the rectangle contained by either of these sides, and the straight line intercepted between the perpendicular let fall upon it from the opposite angle and the acute angle.
Construct a square that shall be equal to the difference between the sum of the squares on two given straight lines and the rectangle under these lines.

3. Through a given point to draw a straight line parallel to a given straight line.
From a given point in the circumference of a circle to draw a chord, when possible, that shall be bisected by a given chord.

4. Find the sum of (1) all the interior angles of any rectilineal figure ; (2) all the exterior angles.
AB, CD the alternate sides of a regular polygon are produced to meet in E, if AC, OE meet in F, O being the centre of the polygon, show that $AF.FO=OF.FE$

6. To divide a given straight line into two parts, so that the rectangle contained by the whole and one of the parts shall be equal to the square on the other part.

If AB be bisected in C and produced to a point D, such that $AC.CD = AD.DB$, then AD is divided in C in the manner required by the proposition.

1. If from any point without a circle two straight lines be drawn, one of which cuts the circle and the other touches it, the rectangle contained by the whole line that cuts the circle and the part of it without the circle shall be equal to the square on the line that touches it.

Any number of circles pass through two given points A and B ; shew that with any given point C in AB produced, as centre, a circle may be described cutting the other circles at right angles, and find its radius.

7. To draw a straight line from a given point either without or in the circumference which shall touch a given circle.

Find the point in the line joining the centres of two circles of different radii, such that if a perpendicular be drawn through it, the tangents to the circles from any point in this perpendicular may be equal.

8. The angle at the centre of a circle is double of the angle at the circumference upon the same base, that is, upon the same part of the circumference.

If a circle be described touching one of the equal sides of an isosceles triangle at the vertex and having the other side as chord, the arc lying between the vertex and base is one-half the arc subtended by the chord.

9. If a straight line touch a given circle and from the point of contact a straight line may be drawn cutting the circle, the angles made by this line with the line touching the circle shall be equal to the angles which are in the alternate segments of the circle.

10. To inscribe an equilateral and equiangular pentagon in a given circle.

If two diagonals of a regular pentagon intersect and a circle be described about the triangle of which the greater segments are two sides, two sides of the pentagon which terminate at the other extremities of these segments are tangents to the circle at these points.

11. To describe a circle about a given square.

Find the relation between the areas of the circles described about and inscribed in a given square.

12. If a straight line be parallel to the base of a triangle it will cut the sides, or the sides produced, proportionally, and if the sides or the sides produced, be cut

proportionally, the straight line which joins **the points**
of section shall be parallel to the base.

13. To find **a** mean proportional between **two given straight**
lines.

JUNIOR AND SENIOR MATRICULATION, 1875.

.*. Junior Matriculants will omit questions 15 and 16, and
Senior Matriculants questions 12 and 13.

1. Define the terms axiom, postulate, scholium, corollory.
2. If two triangles have two sides of the one equal to two
   sides of the other, each to each, but the angle contained by the two sides of the one greater than the
   angle contained by the two sides equal to them, of the
   other, the base of that which has the greater angle
   shall be greater than the base of the other.
3. If a side of any triangle be produced, the exterior angle
   is equal to the two interior and opposite angles; and
   the three interior angles of every triangle are together
   equal to two right angles.
4. Triangles on equal bases and between the same parallels
   are equal to one another.
5. If the square described on one of the sides of a triangle
   be equal to the squares described on the other two
   sides of it, the angle contained by these two sides is a
   right angle.
6. If the diagonals of a quadrilateral bisect each other, it is
   a parallelogram : if the bisecting lines are equal it is
   rectangular ; if the lines bisect at right angles it is
   equilateral.
7. If a straight line be divided into two equal, and also into
   two unequal parts, the squares on the two unequal
   parts are together double of the square on half the
   line and of the square on the line between the points
   of section.
8. Divide a straight line into two parts, so that the rectangle
   contained by the whole and one of the parts may be
   equal to the square on the other part.
9. In the Algebraic solution of the preceding problem, we
   obtain a quadratic equation which gives two values of
   the unknown quantity. Enunciate the Geometrical
   proposition which corresponds to the other root.
10. The sum of the squares on the diagonals of a parallelo-
    gram is equal to the sum of the squares on the sides.
11. The opposite angles of a quadrilateral inscribed in a circle
    are together equal to two right angles.
12. The straight lines bisecting the sides of a triangle at right
    angles meet in a point.

13. Construct a triangle, having given the middle points of sides
14. Describe a circle about a given equilateral and equiangular pentagon.
15. From a given straight line to cut off any part required.
16. Similar triangles are to one another in the duplicate ratic of their homologous sides.

TIME—3 HOURS.

1. Describe an equilateral triangle upon a given finite straight line.

By a method similar to that used in this problem, describe on a given finite straight line an isosceles triangle, the sides of which shall be each equal to twice the base.

2. If a straight line fall on two parallel straight lines, it makes the alternate angles equal to one another, and the exterior angle equal to the interior and opposite angle on the same side; and also the two interior angles on the same side together equal to two right angles.

What objections have been urged against the doctrine of parallel straight lines as it is laid down by Euclid? Where does the difficulty originate and what has been suggested to remove it?

3. In any right angled triangle, the squares described on the sides containing the right angle are together equal to the square of the side subtending the right angle.

Show, by describing a square on the outer side of one side, and on the inner side of the other, that the two squares thus described will cut into *three* pieces, so as exactly to make up the square of the hypotenuse.

4. Divide *algebraically* a given line (*a*) into two parts, such that the rectangle contained by the whole and one part may be equal to the square of the other. Deduce Euclid's construction from one solution and explain the other.

5. If two straight lines within a circle cut one another, the rectangle contained by the segments of one of them is equal to the rectangle contained by the segments of the other.

If, through a point within a circle, two equal straight lines be drawn to the circumference, and produced, they will be at the same distance from the centre.

6. Explain and illustrate the fifth and seventh definitions in the fifth book of Euclid, and shew that a magnitude has a greater ratio to the less of two unequal magnitudes than it has to the greater.

7. With the four lines contain *a*+*b*, *a*+*c*, *a*—*b*, *a*—*c* units respectively, construct a quadrilateral capable of having a circle inscribed in it.

Prove that no parallelogram can be inscribed in a circle except a rectangle ; and that no parallelogram can be described about a circle except a rhomb.

#. Similar triangles are to one another in the duplicate ratio of their homologous sides. How does it appear from Euclid that the duplicate ratio of two magnitudes is the same as that of their squares?

FIRST CLASS PROVINCIAL CERTIFICATES, JULY, 1876.

TIME—THREE HOURS.

N. B.—Algebraic symbols must not be used.

1. (a) The straight line drawn at right angles to the diameter of a circle from the extremity of it, falls without the circle ; and no straight line can be drawn from the extremity, between that straight line and the circumference, so as not to cut the circle. (III 16.)

   (b) Draw a common tangent to two given circles. How many can be drawn ? (*Apollonius.*)

2. (a) The opposite angles of any quadrilateral figure inscribed in a circle are together equal to two right angles. (III 22.)

   (b) If straight lines be drawn from any point on the circumference of a circle perpendicular to the sides of an inscribed triangle, their feet are in the same straight line. (*M. F. Jacobi.*)

3. (a) If the chord of a circle be divided into two segments by a point in the chord or in the chord produced, the rectangle contained by these segments will be equal to the difference of the squares on the radius and on the line joining the given point within the centre of the circle. What propositions in Euclid follow immediately from this?

   (b) Describe a circle which shall pass through a given point and touch two straight lines given in position. (*Apollonius.*)

4. (a) To describe an isosceles triangle, having each of the angles at the base double of the third angle. (IV 10.)

   (b) Construct a triangle having each of the angles at the base equal to seven times the third angle.

5. (a) If the vertical angle of a triangle be bisected by a straight line which also cuts the base, the segments of the base have the same ratio which the other sides of the triangle have to one another ; and, if the segments of the base have the same ratio which the other sides of the triangle have to one another, the straight line

drawn from the vertex to the point of section shall bisect the vertical angle. (VI., 3.)

(*b*) The points in which the bisectors of the external angles of a triangle meet the opposite sides, lie in a straight line.

SECOND CLASS CERTIFICATES, JULY, 1876.

TIME—THREE HOURS.

N B.—Algebraic symbols must not be used. Candidates who take Book II will omit Questions 1, 2 and 3, marked*.

16 *1. The angles at the base of an isosceles triangle are equal to one another ; and if the equal sides be produced, the angles on the other side of the base shall be equal to one another.

8   Where does Euclid require the second part of this theorum?

16 *2. If two triangles have two sides of the one equal to two sides of the other, each to each, but the angle contained by two sides of one of them greater than the angle contained by the two sides equal to them of the other, the base of that which has the greater angle shall be greater than the base of the other.

6   Why the restriction " Of the two sides DE, DF, let DE be the side which is not greater than the other"?

16 *3. If two triangles have two angles of the one equal to two angles of the other, each to each, and have also the sides adjacent to the equal angles in each, equal to one another, then shall the other side be equal, each to each ; and also the third angle of the one to the third angle of the other. (Prove by superposition.)

3   What propositions in Book I are thus proved ?

16 4. If a straight line fall upon two parallel straight lines, it makes the alternate angles equal to one another, and the exterior angle equal to the interior and opposite angle on the same side; and also the two interior angles on the same side together equal to two right angles.

8   What objection may be taken to the twelfth axiom ?

2   What is its converse ?

16 5. In any right-angled triangle, the square which is described on the side subtending the right angle is equal to the squares described on the sides which contain the right angle.

Prove also by dissection and superposition.

Draw through a given point between two straight lines not parallel a straight line which shall be bisected in that point.

The perpendiculars from the angles of a triangle on the opposite sides meet in a point.

Given the lengths of the lines drawn from the angles of a triangle to the points of bisection of the opposite sides, construct the triangle.

If a straight line be divided into two parts, the square on the whole line is equal to the squares on the parts, together with twice the rectangle contained by the parts.

In every triangle, the square on the side subtending an acute angle is less than the squares on the sides containing that angle by twice the rectangle contained by either of these sides, and the straight line intercepted between the perpendicular let fall on it from the opposite angle, and the acute angle.

www.ingramcontent.com/pod-product-compliance
Lightning Source LLC
Chambersburg PA
CBHW021356210326
41599CB00011B/903